# Communications in Computer and Information Science

**720**

*Commenced Publication in 2007*
Founding and Former Series Editors:
Alfredo Cuzzocrea, Xiaoyong Du, Orhun Kara, Ting Liu, Dominik Ślęzak,
and Xiaokang Yang

## Editorial Board

More information about this series at http://www.springer.com/series/7899

Dante Augusto Couto Barone
Eduardo Oliveira Teles
Christian Puhlmann Brackmann (Eds.)

# Computational Neuroscience

First Latin American Workshop, LAWCN 2017
Porto Alegre, Brazil, November 22–24, 2017
Proceedings

Springer

*Editors*
Dante Augusto Couto Barone ⓘ
Federal University of Rio Grande do Sul
Porto Alegre
Brazil

Eduardo Oliveira Teles ⓘ
Federal Institute of Education, Science
  and Technology of Bahia
Camacari
Brazil

Christian Puhlmann Brackmann ⓘ
Federal Institute of Education, Science
  and Technology Farroupilha
Santa Maria
Brazil

ISSN 1865-0929          ISSN 1865-0937 (electronic)
Communications in Computer and Information Science
ISBN 978-3-319-71010-5          ISBN 978-3-319-71011-2 (eBook)
https://doi.org/10.1007/978-3-319-71011-2

Library of Congress Control Number: 2017959594

This Springer imprint is published by Springer Nature
The registered company is Springer International Publishing AG
The registered company address is: Gewerbestrasse 11, 6330 Cham, Switzerland

# Preface

It is with great pleasure that we held LAWCN 2017 (Latin American Workshop on Computational Neuroscience) at the Hotel Plaza São Rafael, Porto Alegre, Brazil, during November 22–24, 2017. The event was organized by the Federal University of Rio Grande do Sul (UFRGS), one of the leading research institutions in South America.

Neuroscience and artificial intelligence have been areas of study with the highest growth in recent years. Together with the solidification of its theoretical bases, and the support of scientific evidence, we are currently experiencing that these two disciplines are strengthening their links increasingly. The biggest source of inspiration for the development of intelligent machines is the brain of human beings.

The first scientific event in the context of computational neuroscience was an initiative that emerged in 2012 as a way of disseminating knowledge and discoveries through discussions between the different areas related to computational neuroscience, mainly at UFRGS and the Rio Grande do Sul state institutions.

For the first time in Brazil, LAWCN 2017 covered recent advances in basic neuroscience, computational neuroscience, neuro-engineering, and artificial intelligence, along with their different interactions. With four renowned international keynote speakers, from the USA, Germany, France, and Mexico, and four reputed national guests, LAWCN intended to consolidate itself as a reference in Latin America, giving the opportunity for interdisciplinary and state-of-art knowledge to be presented as oral presentations.

This volume of Springer's CCIS series, Vol. 720, comprises 15 papers reviewed by 34 members of the Program Committee from 10 different countries worldwide. These articles cover a wide range of issues concerning artificial intelligence and related techniques, including applications in neuroscience and neuro-engineering. Showcasing diversity and quality, these papers point to future directions of research and developments in this exciting area.

The success of the workshop was due to the dedication and support of many individuals and organizations as well as the funding from CAPES (Brazilian Coordination for the Improvement of Higher Education Personnel) and IBRO (International Brain Research Organization). On behalf of the Organizing Committee, we would like to thank all authors for careful preparation of their papers, and all speakers for sharing their work, experience, and insights at the workshop. All full papers submitted were reviewed by at least two members of the Program Committee. We are grateful to all of them for their important contributions to the workshop.

We hope that the 2017 edition of LAWCN offered good opportunities motivation for important contacts and sharing of knowledge and experiences, which will motivate its participants and readers of these proceedings to foster their careers in all fields linked to computational neuroscience, giving support and encouragement for the upcoming LAWCN editions.

October 2017

Dante Augusto Barone
Eduardo Oliveira Teles
Christian Brackmann

# Organization

## Latin-American Workshop on Computational Neuroscience 2017

November 22–24, 2017
Porto Alegre, Brazil

### Organized by

Federal University of Rio Grande do Sul, Brazil

### Co-organized by

Federal Institute of Education, Science and Technology Farroupilha
Federal Institute of Education, Science and Technology of Bahia

### Conference Chairs

| | |
|---|---|
| Dante Augusto Couto Barone | UFRGS, Brazil |
| Marino Bianchin | UFRGS, Brazil |

### Organizing Committee

| | |
|---|---|
| Christian Puhlmann Brackmann | IFFAR, Brazil |
| Dante Augusto Couto Barone | UFRGS, Brazil |
| Eduardo Oliveira Teles | IFBA, Brazil |
| Edison Pignaton de Freitas | UFRGS, Brazil |
| Jaime Andres Riascos Salas | UFRGS, Brazil |
| Marino Muxfeldt Bianchin | UFRGS, Brazil |
| Paulo Rogério de Almeida Ribeiro | UFMA, Brazil |
| Pedro Schestatsky | HCPA, Brazil |
| Rodrigo Alejandro Sierra | UFRGS, Brazil |

### Program Committee Chair

| | |
|---|---|
| Dante Augusto Couto Barone | UFRGS, Brazil (Chair) |

### Program Committee

| | |
|---|---|
| Alex Barradas | UFMA, Brazil |
| Alexandre César Muniz de Oliveira | UFMA, Brazil |

# Contents

**Neural Networks**

A Network of Spiking Neurons Performing a Relational
Categorization Task.................................................. 3
   *Lucas Ferreira Alves, Fernando Lopes Araujo Junior,*
   *Bruno Andre Santos, and Rogerio Martins Gomes*

Chaotic Synchronization of Neural Networks in FPGA ................ 17
   *Elias de Almeida Ramos, Vitor Bandeira, Ricardo Reis,*
   *and Guilherme Bontorin*

Assessing the Performance of Convolutional Neural Networks
on Classifying Disorders in Apple Tree Leaves..................... 31
   *Pedro Ballester, Ulisses B. Correa, Marco Birck, and Ricardo Araujo*

Towards Graffiti Classification in Weakly Labeled Images
Using Convolutional Neural Networks.............................. 39
   *Glauco R. Munsberg, Pedro Ballester, Marco F. Birck,*
   *Ulisses B. Correa, Virginia O. Andersson, and Ricardo M. Araujo*

Computational Models for the Propagation of Spreading
Depression Waves................................................. 49
   *Guillem Via, Jean Faber, and Esper Abrão Cavalheiro*

**Artificial Intelligence**

How Artificial Intelligence is Supporting Neuroscience Research:
A Discussion About Foundations, Methods and Applications ........... 63
   *Rafael T. Gonzalez, Jaime A. Riascos, and Dante A.C. Barone*

**Computer Vision**

Investigating Crime Rate Prediction Using Street-Level Images
and Siamese Convolutional Neural Networks ....................... 81
   *Virginia O. Andersson, Marco A.F. Birck, and Ricardo M. Araujo*

Developing of a Video-Based Model for UAV Autonomous Navigation .... 94
   *Wilbert G. Aguilar, Vinicio S. Salcedo, David S. Sandoval,*
   *and Bryan Cobeña*

## Machine Learning

Cyberbullying Classification Using Extreme Learning Machine
Applied to Portuguese Language........................... 109
   *Jim Jones da Silveira Marciano,*
   *Eduardo Mazoni Andrade Marçal Mendes,*
   *and Márcio Falcão Santos Barroso*

Pseudorehearsal Approach for Incremental Learning of Deep
Convolutional Neural Networks ........................... 118
   *Diego Mellado, Carolina Saavedra, Stéren Chabert, and Rodrigo Salas*

## Graphics Systems and Interfaces

A Brain Computer Interface Using VEP and MMSC for Driving
a Mechanical Arm......................................... 129
   *Marcos Antônio Abdalla Júnior, Carlos Alberto Cimini Júnior,*
   *Márcio Falcão Santos Barroso, and Leonardo Bonato Félix*

An Adaptive User Interface Based on Psychological Test
and Task-Relevance....................................... 143
   *Jaime A. Riascos, Luciana P. Nedel, and Dante C. Barone*

## Decision Trees

Decision Tree to Analyses EEG Signal: A Case Study
Using Spatial Activities ................................. 159
   *Narúsci dos Santos Bastos, Diana Francisca Adamatti,*
   *and Cleo Zanella Billa*

## Nonlinear Equations

Multi-Network-Feedback-Error-Learning with Automatic Insertion:
Validation to a Nonlinear System ........................ 173
   *Alex N.V. Santos, Paulo R.A. Ribeiro, Areolino de Almeida Neto,*
   *and Alexandre C.M. Oliveira*

## Nanoelectromechanical Systems

Planned Obsolescence Using Nanotechnology for Protection
Against Artificial Intelligence........................... 189
   *Mirkos Ortiz Martins and Dante Barone*

**Author Index** ........................................ 195

# Neural Networks

# A Network of Spiking Neurons Performing a Relational Categorization Task

Lucas Ferreira Alves[(⊠)] [iD], Fernando Lopes Araujo Junior[iD],
Bruno Andre Santos[iD], and Rogerio Martins Gomes[iD]

Department of Computer Engineering, CEFET-MG, Belo Horizonte, MG, Brazil
ferreira.alves@live.com, fer.ajnr@gmail.com, bruno@decom.cefetmg.br,
rogerio@lsi.cefetmg.br

**Abstract.** The study of the storage and transmission of information in neural networks is an important and challenging field of research. Studies in this area aim to understand the process by which the neural systems encode and process information originating from the environment. This work aims to develop a computational model that can be used to study how neural systems encode and relate information about external stimuli. To fulfill this purpose, a computational model composed by a spiking neuron network is developed to perform a task of relational categorization that consists in measuring the relation between the intensities of two signals applied to network. A Genetic Algorithm is used to optimize the synaptic weights of the network. The results show that the network is able to perform the task of relational categorization according to a threshold defined as error rate, as well as shows that the ability of the network to detect the relation between the signals depends on the minimum and maximum difference in the number of spikes in a given time window.

**Keywords:** Spiking neurons · Sensory processing · Network dynamics

## 1 Introduction

The study of storage and transmission of information in neural networks is a broad research field called "neural coding" [9,11,22,24]. Studies in this area aim to understand the process by which neural systems encode and process information originated from the environment and generate primitive motor functions and even complex cognitive phenomena [2,20,21].

Researches in this field attempted to show that the information can be stored in different ways, such as: in the spiking frequency of the neurons [1,18], in the temporal relation between the first neuronal spiking and the stimulus time [8], in the time between the neuronal spikings [17] and in the spatial distribution of local synchronization of neural oscillation [12,23]. Although there is a wide variety of empirical results, there is no consensus on neural coding.

When the neural coding is studied, researchers seek to understand how the intrinsic properties of external stimuli (*e.g.* the color of an object [5], the gaps

© Springer International Publishing AG 2017
D.A.C. Barone et al. (Eds.): LAWCN 2017, CCIS 720, pp. 3–16, 2017.
https://doi.org/10.1007/978-3-319-71011-2_1

between two sounds [6] and the intensity of a light source [3]) are encoded by neural systems. The aim of our work is to build a computational model from which we can study how the intensity of an abstract input signal is encoded by a spiking neural network.

Most computational models developed to study the neural coding are designed with the assumption that the information is encoded by the neural network according to predefined mechanisms. For example, Yu and colleagues [27] developed a spiking neural network to perform various classification tasks by assuming that the information is encoded by the timing of the neural response relative to the stimulus onset. They demonstrated that this type of temporal coding is a viable mechanism for fast neural information processing. Bucci and colleagues [4] developed a neuromorphic robot controlled by a spiking neural network capable of classifying a person's hand movements. They modeled two types of networks by using unsupervised learning techniques. In each type of network they assumed that the information was encoded by spiking rates and by time-locked but not synchronous firing patterns (*i.e.* polychronous groups [16]), respectively. They found that both encoding mechanisms were capable of classifying the hand movements, but the temporal coding outperformed the firing rate coding.

In this work, there is no assumption about the neural mechanism that encodes the information. It is used an optimization algorithm to adjust the parameters of the model in order to obtain a network that stores the information about the intensity of external input signals. This approach allows us to find out different and efficient encoding mechanisms which could, for example, encompass a combination of rate and temporal coding.

In order to develop a network that encodes the information about the intensity of input signals, a relational categorization task is implemented. In this task, two signals are presented to the network in different time windows. Based on the relation between their intensities, the network should generate an output indicating whether the first signal is higher or lower than the second one. Notice that the network should store intrinsic properties of the input signals (*i.e.* their intensities) so that they can be related in order to generate the correct output. Differently from the traditional classification problems where objects are grouped into categories, in the relational categorization task the network has to identify relational structures (*e.g.* higher and lower) among objects. This type of task has been investigated in cognitive science from a conceptual perspective [7] and also by using computational models based on prior assumptions about the phenomenon [13, 25].

A computational model which has inspired this work was developed by Willians and Beer [26]. In this model, a simulated robot controlled by a Continuous Time Recurrent Network (CTRNN) was evolved to perform the relational task. They found out that the information about the external stimuli was stored in the activation of a specific neuron and in the time at which a bifurcation in system dynamics occurs. In our work, we use a spiking neuron network, rather than a CTRNN, to perform the relational task. The advantage of using this type of neuron is that it allows us to study the network dynamics in terms of spikes, which could give us insights about the spiking dynamics underlying the relational categorization task.

The methods used to develop the model are presented in Sect. 2 and the results in Sect. 3.

## 2  Methods

The relational categorization task implemented is described in Sect. 2.1. The neuron and the network models are presented in Sects. 2.2 and 2.3, respectively, and the method used the adjust the parameters of the model is presented in Sect. 2.4.

### 2.1  Relational Categorization Task

A detailed timeline of the experiment is shown in Fig. 1. During a trial of the experiment, two inputs of distinct intensities are applied to an input neuron. The first input (referred to as $i_1$) is applied during a time window $T1$ [25, 75] ms. During a time window $T2$ [75, 100] ms, no input is applied to the input neuron. The second input (referred to as $i_2$) is applied to the same input neuron during a time window $T3$ [100, 150] ms. The analysis of the output begins after $i_2$ is removed and continues for 400 ms. In the analysis of the output, the number of spikes of a specific neuron is counted. In successful trials of the task, a higher number of spikes in the output neuron means that $i_1 > i_2$ and a lower number of spikes means that $i_1 < i_2$. In total, a trial of the experiment lasts 550 ms. Notice that in order to store the information about $i_1$ the network should maintain its activity during $T_2$, even without an input signal.

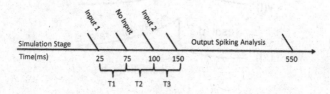

**Fig. 1.** Timeline for the experiment. The first input ($i_1$) is applied to an input neuron during 50 ms (T1 = [25, 75] ms). The network does not receive input for 25 ms (T2 = [75, 100] ms). A second input $i_2$ that lasts 50 ms is applied to the input neuron during T3 = [100, 150] ms. After T3, the number of spikes from an output neuron is counted during a time window of 400 ms.

## 2.2   Neuron Model

In this work, the Izhikevich neuron model is used to implement the spiking neural network. The Izhikevich model presents a balance between biological plausibility and computational viability [14, 15]. It does not take into account the neural biological structures, but is capable of reproducing the neural membrane potential dynamics with great plausibility. This neuron model is described by Eqs. 1 and 2 with an auxiliary after-spike resetting represented by Eq. 3:

$$v' = 0.04v^2 + 5v + 140 - u + I, \tag{1}$$

$$u' = a(bv - u) \tag{2}$$

$$\text{if } v \geq 30 \text{ mV, then } \begin{cases} v \leftarrow c \\ u \leftarrow u + d \end{cases} \tag{3}$$

where $v$ is the membrane potential of the neuron, $u$ is the membrane recovery variable, $I$ is the current received from external stimuli and from other neurons, $a$ is the timescale of the recovery variable $u$ and $b$ the sensitivity of $u$ for subthreshold fluctuations of the membrane potential. Variables $c$ and $d$ represent the after-spike reset of the membrane potential $v$ and the recovery variable $u$, respectively. The influence of the parameters $a, b, c, d$ on the dynamics of $v$ and $u$ is illustrated in Fig. 2.

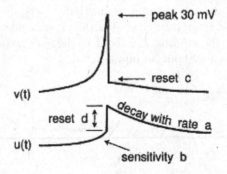

**Fig. 2.** Illustration of how the dynamics of $v$ and $u$ are affected by the parameters $a$, $b$, $c$ and $d$ [15].

Different types of neurons can be simulated by assigning specific values to the parameters $a$, $b$, $c$ and $d$ [14, 15]. In this work, the following values have been used: $a = 0.02$, $b = -0.1$, $c = -55$ and $d = 6$ (values taken from [15]). By using these values, the input neuron can encode the strength of a continuous input signal in its firing rate (from 2 Hz to 200 Hz). The other neurons, which receive spikes as input, act as integrators of the input current and can be used to detect coincident or nearly coincident spikes.

## 2.3 Network Model

Spiking neural networks of $n$ neurons were implemented. A neuron, referred to as neuron 0, was selected to receive the input signals. The other $n-1$ neurons were implemented as a fully connected network. An illustration of the neuron connections can be seen in Fig. 3. Neuron 0 does not receive post-synaptic signals from the network. Another neuron, referred to as neuron 5, was selected to represent the output of the network. The number of spikes of this neuron was analyzed in order to check whether the network has correctly performed the comparison between the input signals.

The connection weights of the neurons were defined by a matrix W where the lines represent the indexes of the presynaptic neurons, and the columns represent the indexes of the postsynaptic neurons. Values for the weights were adjusted by a genetic algorithm (explained later).

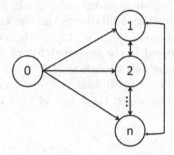

**Fig. 3.** Neuron connections representation: Neuron 0 acts as the network input and has an one way connection to the remaining neurons. Neurons from 1 to $n$ form a fully connected network.

The input range applied to neuron 0 was defined after testing the frequency output of an individual neuron. It was observed that the neuron spiking frequency is nearly linear for inputs in the range [3, 33]. For greater values, the output frequency presented some plateaus which are not feasible for detecting small difference between the input signals.

The input current applied to the other neurons is described by Eq. 4.

$$I_i = \sum_{j}^{n} W_{ij} g(t)_j, \tag{4}$$

where $I$ is the input current for each neuron $i$ (except neuron 0), $W$ is the synaptic weight of the connection between the neurons i and j and $g$ is a conductance model representing the synaptic dynamics (see Eq. 5).

$$g(t) = \frac{t}{\tau} e^{1-\frac{t}{\tau}}, \tag{5}$$

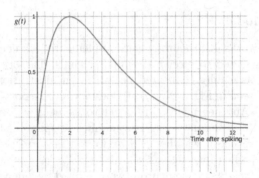

**Fig. 4.** Alpha function over time for a conductance coefficient equals 2. Neuron connection weights are multiplied by the value of g(t).

where $t$ is the time after a neuron spike, and $\tau$ is the conductance coefficient. The behavior of the g(t) function is illustrated in Fig. 4.

Equation 5 is known as alpha-function [19]. Postsynaptic conductance is an important mechanism involved in the sustainability of network information, as well as in its noise robustness and synchrony [19].

In order to obtain a network performing the relational categorization task, a genetic algorithm was used to adjust the values of the connection weight matrix $W_{i,j}$.

### 2.4   Parameters Optimization

In a successful trial of the experiment, the network should present different number of spikes in the output neuron depending on the input relations $i_1 > i_2$ and $i_1 < i_2$. The Microbial Genetic Algorithm [10] was used to optimize the synaptic weights between neurons. An initial population of 30 networks was initialized with random neuron connection weights.

The fitness of each network is calculated as follows. After selecting a network to evaluate, two values of inputs are randomly chosen. The lowest value is attributed to $i_1$, and the highest one to $i_2$ (*i.e.* $i_1 < i_2$). A trial of the experiment is ran and the number of spikes is counted and stored in a variable $s_{lh}$ (spikes for $i_1 = low$ and $i_2 = high$). Then, the values of $i_1$ and $i_2$ are exchanged (*i.e.* $i_1 > i_2$), a new trial is ran and the number of spikes is counted and stored in another variable $s_{hl}$ (spikes for $i_1 = high$ and $i_2 = low$). A partial fitness of the network for the pair of inputs $i_1$ and $i_2$ is calculated as: $F_{i_1,i_2} = s_{lh} - s_{hl}$. The network is evaluated for a hundred randomly chosen values of $i_1$ and $i_2$ and its final fitness is given by the mean of the partial fitnesses. The fittest network is the one with the greatest final fitness. Notice that the fitness function tries to maximize the the difference in the number of spikes between the trials where $i_1 < i_2$ and $i_1 > i_2$.

In the genetic algorithm, the chromosomes of the losing network (lower fitness) are recombined with the winning network (higher fitness) at a rate of 0.6 and mutated at a rate of 0.05.

While the connection weights $W_{i,j}$ were optimized by the genetic algorithm, the minimum and maximum values $W_{i,j}$ (*i.e* its search space), the network size ($n$) and the conductance coefficient ($\tau$) were adjusted by following an empirical approach. The values of $W_{i,j}$, $n$ and $\tau$ influenced the ability of the network to perform the relational task. Some notable examples were small networks with low values for the connection weights that did not present any activity during the analysis period. In contrast, large networks with high connection weights became over-saturated and presented activity regardless of the given input. Higher values of $\tau$ were important to keep the network activity during the time window $T2$ and after removing $i_2$. The values empirically found for these parameters were the following: network size ($n = 10$), conductance coefficient($\tau = 5$) and connection weight search space (0–15).

The next section presents the analysis of the fittest network evolved by the genetic algorithm.

## 3    Results

Figure 5 shows how the best fitness of the population changes over evolution process. Taking into account the objective function, the increase in fitness represents a greater distinction between the output patterns in cases where $i_1 > i_2$ and $i_2 > i_1$.

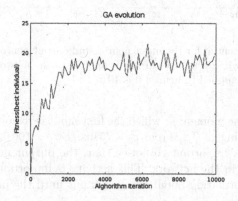

**Fig. 5.** Optimization process: fitness of the best individual of the population.

The following analysis is based on the best network evolved by the genetic algorithm. The dynamics of the network for a single trial of the experiment and the network performance for all possible combinations of inputs are shown in Sects. 3.1 and 3.2, respectively.

## 3.1   Neural Dynamics

In the first part of the dynamical analysis, it is presented the neural activation during a single trial of the experiment where the first input is equal to 10 and the second input is equal to 20. Then, it is presented a more general analysis of the network capacity to perform the relational categorization task for input signals within the range [3, 33].

Figures 6, 7 and 8 show the membrane potential for the input and output neurons during a trial of the relational categorization task. Figure 6 shows the time at which the first input is applied to the input neuron (neuron 0) at $t = 25$ ms, which defines the beginning of T1. Notice that, the information about the input arrives at the output neuron (neuron 5) at $\sim 30$ ms, which is represented by an increase in the neuron membrane potential.

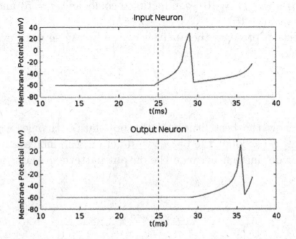

**Fig. 6.** Membrane potential for the input neuron and output neurons. It is possible to observe the moment where the first current is applied to the network ($t = 25$). In this example, the current applied to neuron 0 is 10.

Figure 7 shows the moment at which the first input is removed (at $t = 75$ ms), defining the beginning of T2. From $t = 75$ ms to $t = 100$ ms, the network is not stimulated. It is important to notice that, the output neuron keeps firing even without input in the network. This activity is important for the network to maintain the information about the first input until the presentation of the second input at T3.

The transition from T2 to T3 and from T3 to the analysis period can be seen in Fig. 8. Note that after application of the second input ($i_2$) the input neuron still fires a spike after the removal of the current at $t = 150$ ms. This spike occurs because the neuron membrane potential had already crossed the spiking threshold when the input current was removed.

The difference between the number of spikes for the cases where $i_1 > i_2$ and $i_1 < i_2$ can be seen in Figs. 9 and 10, respectively. The difference between the

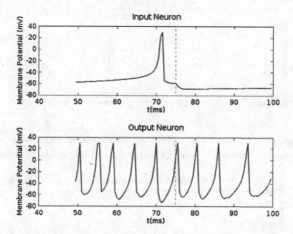

**Fig. 7.** Membrane potential for neurons (0) and (5). The first input is applied until $t = 75$ ms, as shown by the dashed line. Neuron 5 continues to spike during T2 even without input current.

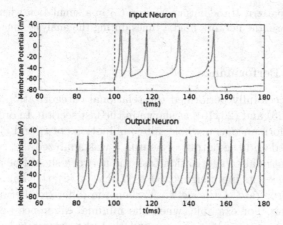

**Fig. 8.** Membrane potential for neurons (0) and (5). The current is applied to neuron 0 from $t = 100$ ms to $t = 150$ ms. Neuron 0 leaves its stable state at $t = 100$ ms. In this example the current applied to neuron 0 is 20.

output patterns for the two cases lies in the time it takes for the network activity to disappear. When $i_1 > i_2$ the network takes a longer time to stop firing than when $i_1 < i_2$.

In a successful trial of the task the number of spikes in the output neuron, in the time window [150, 550] ms, is within [1, 5] for $i_1 < i_2$ and within [30, 35] for $i_1 > i_2$. These intervals were obtained by the genetic algorithm.

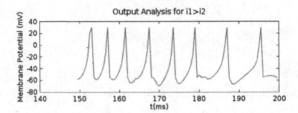

**Fig. 9.** Spiking pattern of the output Neuron (5) in a simulation where $i_1 = 20$ and $i_2 = 10$. In this case, the network fired 33 times during the analysis time window.

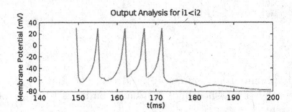

**Fig. 10.** Spiking pattern of the output Neuron (5) in a simulation where $i_1 = 10$ and $i_2 = 20$. In this case, the network fired 4 times during the analysis time window.

### 3.2 Network Performance

The network successfully performed the relational categorization task for both input pairs (10, 20) and (20, 10), as shown in the last section. In order to analyze the network performance as a whole, all combinations of $i_1$ and $i_2$ in the range [3, 33] were tested and the output of neuron 5 was analyzed.

As in the relational categorization task the network should be able to detect the relation between its input signals ($i1 > i2$ or $i1 < i2$), the aim of the analysis is to find the error rate of the network when the difference between $i_1$ and $i_2$ ($i_1 - i_2$) increases. For example, when the minimal difference between $i_1$ and $i_2$ is 5 (*i.e.*, $i_1 - i_2 >= 5$), it is expected that the error rate will be greater than when the minimal difference is 10 (*i.e.*, $i_1 - i_2 >= 10$). The value of the difference minimal between $i_1$ and $i_2$ will be referred to as *threshold*. Figure 11 shows the error rate of the network for different values of the threshold. For easier visualization, the absolute value of the difference $i_1 - i_2$ was used.

When the difference between $i_1$ and $i_2$ is equal or greater than 5 ($abs(i_1 - i_2) >= 5$), the error rate is 0.25 for $i_1 > i_2$ (green line) and 0.14 for $i_1 < i_2$ (red line). The error rate for $i_1 > i_2$ (green line) is bigger than the error rate for $i_1 < i_2$ (red line) for all values of $abs(i_1 - i_2) < 11.5$. If it is desired an error rate smaller than 10% (0.1 in the y-axis), the threshold should be greater than 10 (*i.e.* $abs(i_1 - i_2) >= 10$); and if it is desired an error rate smaller than 5% (0.05 in the y-axis), the threshold should be greater than 17.5 (*i.e.* $abs(i_1 - i_2) >= 17.5$). In order to get an error rate equal 0, for $i_1 > i_2$ the threshold should be 19; and for $i_1 < i_2$ the threshold should be 28.

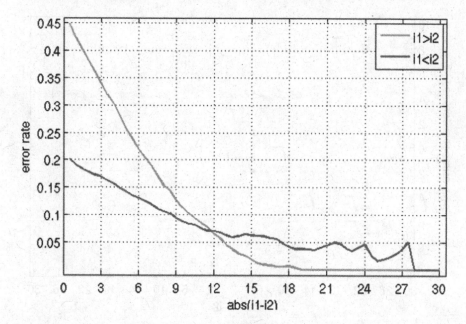

**Fig. 11.** Error rate in the output of the network. The error rate is shown in the y-axis and the absolute value of the difference between $i_1$ and $i_2$ $(abs(i_1 - i_2))$ is shown in the x-axis. The green and red lines show the cases where $i_1 > i_2$ and $i_1 < i_2$, respectively (see legend). Values of $i_1$ and $i_2$ varies in the range [3,33] with step of 0.5, that is why their difference (shown in the x-axis) changes from 0 to 30. The cases where $i_1 = i_2$ (*i.e.* $abs(i_1 - i_2) = 0$) were not considered. (Color figure online)

When the difference between the input signals is low, the error rate of the network is high because the number of spikes generated by each signal is similar. Figure 12 shows the difference in the spike numbers generated by all values of $i_1 - i_2$ (from $-30$ to $30$). For example, when the difference between the inputs is equal to 5 (*e.g.* $i_1 = 25$ and $i_2 = 20$ or $i_1 = 11$ and $i_2 = 6$), the difference between the number of spikes generated by each input signal can be equal to 2, 3 or 4, depending on the values of $i_1$ and $i_2$. Just as a reference, the maximum number of spikes generated by an input signal is 9 for $i = 33$ (value not shown in the graphics).

If it is desired an error rate smaller than 0.1 for $i_1 > i_2$, then $i_1 - i_2$ should be greater than 10 (values taken from Fig. 11). When $i_1 - i_2 = 10$, the difference in the number of spikes can be 4, 5 or 6 (values taken from Fig. 12). In other words, when the difference in the number of spikes is greater or equal than 4, the error rate is smaller than 0.1 for the cases where $i_1 > i_2$.

When $i_1 - i_2 = 13$, the difference in the number of spikes can be from 4 to 7 (values taken from Fig. 12). The error rate is 0.05 for the difference $abs(i_1 - i_2) = 13$ and $i_1 > i_2$ (values taken from Fig. 11). On the other hand, for $i_1 < i_2$ the same error rate is obtained when $abs(i_1 - i_2) = 17.5$, which gives a spike number difference of $-3$, $-4$ or $-5$ (values taken from Fig. 12). In other words, in order

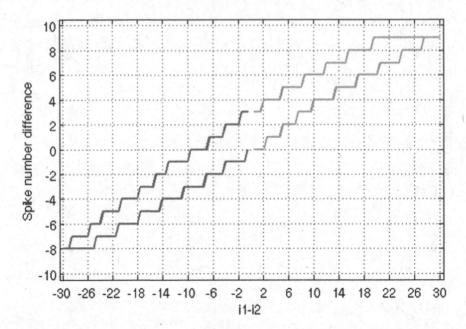

**Fig. 12.** The x-axis shows the difference in the input signals $(i_1 - i_2)$ and the y-axis shows the number of spikes generated by $i_1$ minus the number of spikes generated by $i_2$. The green lines (red lines) show the differences in the number of spikes for $i_1 > i_2$ $(i_1 < i_2)$. Different combinations of $i_1$ and $i_2$ that generate the same value for $i_1 - i_2$ produce a different spike number difference (represented by the area within the lines). (Color figure online)

to get an error rate of 0.05, the spike number difference should be within [4, 7] when $i_1 > i_2$ and within [3, 5] (in absolute value) when $i_1 < i_2$. This shows that there is no symmetry in the detecting the relations $i_1 > i_2$ and $i_1 < i_2$.

## 4   Discussion and Conclusion

The network was able to perform the relational categorization task according to the error rate presented in Fig. 11. The capability of the network to detect the relation between the signals depends on the minimum and maximum difference in the number of spikes within T1 and T3 (time windows where the input signals are presented). In order to try to obtain a smaller error rate, the length of these time windows could be increased or the maximum frequency of the Izhikevich neuron could be increased by changing the parameters $a$ and $b$ (Eq. 2). Higher values of $a$ and $b$ can make a neuron to converge to its stable state faster and consequently fire more action potentials. For both cases, the parameters of the model should be optimized again.

The error rate could also be decreased by increasing the time window T2 between the application of the input signals $i_1$ and $i_2$. When the input $i_2$ is applied, the membrane potential of the input neuron is still recovering from the last spike generated by $i_1$. The closer the membrane potential is to its resting state, the easier it is for the neuron to fire a spike. For this reason, the same values of input signals (e.g. $i_1 = 15$ and $i_2 = 15$) do not generate the same number of spikes during T1 and T3. By increasing the length of T2, the input neuron will have more time to achieve its resting state and, consequently, could fire the same number of spikes for the same values of $i_1$ and $i_2$. We expect that this symmetry in the number of spikes for the same input signals could decrease the error rate of the network.

Notice that the results presented are valid only in the input interval analyzed here [3,33]. Besides, when it is said that the difference between the input signals should be greater than a threshold ($abs(i_1 - i_2) > threshold$), it is not guaranteed that it can be greater than 30 (33-3). The reason is that the number of spikes within the time windows T1 and T3 are limited by the size of these windows. Another reason is that the frequency of the neuron does not change linearly with its input, it saturates at high frequency showing some plateaus (not studied here).

As mentioned, this work is a first step towards the understanding of how sensory stimulation can be encoded by spiking neural networks. Subsequent studies will focus on the analysis of how the information was stored and processed by the network. Particularly, we are interested in understanding how the network stores the information about the first input so that it can be related with the information about the intensity of the second input which is presented few milliseconds later.

**Acknowledgment.** The authors thank the support of CAPES-Brazil, CNPq-Brazil, FAPEMIG, and CEFET-MG.

# References

1. Adrian, E.D.: The impulses produced by sensory nerve endings. J. Physiol. **61**(1), 49–72 (1926)
2. Aguilera, M., Bedia, M.G., Santos, B.A., Barandiaran, X.E.: The situated HKB model: how sensorimotor spatial coupling can alter oscillatory brain dynamics. Front. Comput. Neurosci. **7**, 117 (2013)
3. Barlow Jr., R.B.: Neural coding of light intensity. In: Ratio Scaling of Psychological Magnitude: In Honor of the Memory of SS Stevens, p. 163 (2013)
4. Bucci, L.D., Chou, T.S., Krichmar, J.L.: Sensory decoding in a tactile, interactive neurorobot. In: 2014 IEEE International Conference on Robotics and Automation (ICRA), pp. 1909–1914. IEEE (2014)
5. De Valois, R.L., De Valois, K.D.: Neural Coding of Color (1997)
6. Frisina, R.D.: Subcortical neural coding mechanisms for auditory temporal processing. Hear. Res. **158**(1), 1–27 (2001)
7. Goldwater, M.B., Schalk, L.: Relational categories as a bridge between cognitive and educational research. Psychol. Bull. **142**(7), 729–757 (2016)
8. Gollisch, T., Meister, M.: Rapid neural coding in the retina with relative spike latencies. Science **319**(5866), 1108–1111 (2008)

9.  Grodzinsky, Y., Nelken, I.: The neural code that makes us human. Science **343**(6174), 978–979 (2014)
10. Harvey, I.: The microbial genetic algorithm. In: Kampis, G., Karsai, I., Szathmáry, E. (eds.) ECAL 2009. LNCS, vol. 5778, pp. 126–133. Springer, Heidelberg (2011). https://doi.org/10.1007/978-3-642-21314-4_16
11. Haxby, J.V., Connolly, A.C., Guntupalli, J.S.: Decoding neural representational spaces using multivariate pattern analysis. Annu. Rev. Neurosci. **37**, 435–456 (2014)
12. Hipp, J.F., Engel, A.K., Siegel, M.: Oscillatory synchronization in large-scale cortical networks predicts perception. Neuron **69**(2), 387–396 (2011)
13. Hummel, J.E., Holyoak, K.J.: Distributed representations of structure: a theory of analogical access and mapping. Psychol. Rev. **104**(3), 427 (1997)
14. Izhikevich, E.M.: Simple model of spiking neurons. IEEE Trans. Neural Netw. **14**(6), 1569–1572 (2003)
15. Izhikevich, E.M.: Which model to use for cortical spiking neurons? IEEE Trans. Neural Netw. **15**(5), 1063–1070 (2004)
16. Izhikevich, E.M.: Polychronization: computation with spikes. Neural Comput. **18**(2), 245–282 (2006)
17. Reinagel, P., Reid, R.C.: Precise firing events are conserved across neurons. J. Neurosci. **22**(16), 6837–6841 (2002)
18. Rieke, F., Warland, D., van Steveninck, R.D.R., Bialek, W.: Spikes: Exploring the Neural Code. MIT press (1999)
19. Roth, A., van Rossum, M.: Computational Modeling Methods for Neuroscientists (2009)
20. Santos, B., Barandiaran, X., Husbands, P., Aguilera, M., Bedia, M.: Sensorimotor coordination and metastability in a situated HKB model. Conn. Sci. **24**(4), 143–161 (2012)
21. Santos, B.A., Barandiaran, X.E., Husbands, P.: Synchrony and phase relation dynamics underlying sensorimotor coordination. Adapt. Behav. **20**(5), 321–336 (2012)
22. Seth, A.K.: Neural coding: rate and time codes work together. Curr. Biol. **25**(3), R110–R113 (2015)
23. Singer, W.: Dynamic formation of functional networks by synchronization. Neuron **69**(2), 191–193 (2011)
24. Stanley, G.B.: Reading and writing the neural code. Nat. Neurosci. **16**(3), 259–263 (2013)
25. Tomlinson, M.T., Love, B.C.: From pigeons to humans: grounding relational learning in concrete examples. In: Proceedings of the National Conference on Artificial Intelligence, vol. 21, p. 199. AAAI Press/MIT Press, Menlo Park/Cambridge/London (1999, 2006)
26. Williams, P.L., Beer, R.D., Gasser, M.: An embodied dynamical approach to relational categorization. In: Proceedings of the Cognitive Science Society, vol. 30 (2008)
27. Yu, Q., Tang, H., Tan, K.C., Yu, H.: A brain-inspired spiking neural network model with temporal encoding and learning. Neurocomputing **138**, 3–13 (2014)

# Chaotic Synchronization of Neural Networks in FPGA

Elias de Almeida Ramos[✉], Vitor Bandeira, Ricardo Reis,
and Guilherme Bontorin

Institute of Informatics, PPGC/PGMicro, Federal University of Rio Grande do
Sul (UFRGS), Mailbox 15.064, Porto Alegre, RS 91.501-970, Brazil
{elias.ramos, vvbandeira, reis}@inf.ufrgs.br,
bontorin@ufpr.br

**Abstract.** The objective of this work is to obtain a complete synchronization of Hopfield Neural Networks (HNN) with a delay using a Field Programmable Gate Array (FPGA) simulating in real-time a Natural Neural Networks (NNN). This work is motivated by research in Neurosciences involving the implantation of chips between the skull and the brain to prevent or ameliorate diseases such as Parkinson's, Epilepsy and Depression. Our contribution is the introduction of new synchronization techniques based on the Qualitative Theory of Differential Equations, Chaos Theory and Algebraic Topology substituting calculations using the Lyapunov Stability Criterion (LSC). The presented technique does not depend on the Neural Networks to be synchronized but also presents a lower computational cost in comparison with previous works. The results show that FPGAs are good platforms for such experiments.

**Keywords:** FPGA · Neurosciences · Synchronization · Neural Networks · Chaos Theory

## 1 Introduction

The brain is a system composed of neurons arranged in a highly complex network. In biological studies, it was observed that isolated neurons can emit signals irregularly, but analyzing a network, the signals are synchronized over time [1]. Synchronization is fundamental in certain stages of sleep, in addition to be related to some pathology such as Parkinson's disease [2]. In recent years, brain implants have emerged as an alternative to the usual medications for treating neurological diseases. Brain implants are electronic devices connected directly to the brain, usually on the surface below the skull. Implants aim to block, record or stimulate signals from networks of neurons or even isolated neurons.

It is of scientific knowledge that diseases of cerebral origin originate by the deficiency of the neurons in the sending and receiving signals. Epilepsy, Parkinson's disease, and depression problems are being treated with Deep Brain Stimulation [2] which consists of the stimulation of isolated neurons or whole deficient regions. Among the current deficiencies, we can highlight synchronization failures of the neurons. Following research in Computational Neuroscience [3–5], information

D.A.C. Barone et al. (Eds.): LAWCN 2017, CCIS 720, pp. 17–30, 2017.
https://doi.org/10.1007/978-3-319-71011-2_2

processing in the brain is modeled by chaotic systems, more precisely, by nonlinear differential equations (Dynamical Systems).

In the study of the brain, the last decades show that it is necessary to introduce new techniques as nonlinear dynamics, since nerve cell activities or pattern recognition can't be totally modeled by classical tools [5]. In addition, experiments show that small disturbances in neural activities generate significant changes, the so-called "Butterfly Effect" [41].

In the last 25 years, techniques for Dynamic System Synchronization have been developed [11–13]. Such synchronizations have also application in several areas of computing, such as Cryptography and Signal Processing [15, 16]. On the other hand, to simulate Natural Neural Networks (NNN) we use Artificial Neural Networks (ANN). They are computational models inspired by the central nervous system of an animal where they are widely applied in problems of a computational nature. ANNs has evolved with robust techniques for solving machine learning problems, and pattern recognition [17, 18]. Among the models of ANNs, we can cite the Hopfield Neural Networks (HNN), which employs a principle called the storage of information in the form of dynamically unstable attractors [18]. The information retrieval happens through a dynamic process of updating the states of the neurons where the neuron to be updated is chosen randomly. Both the ANNs and synchronization problems can use parallel processing [18]. We are using a Field Programmable Gate Array (FPGAs). They are truly parallel because different processing operations do not have to struggle for the same resources. This allows users to create multiple specific task cores, possibly processed in parallel using multiples cores from a same chip. Hardware execution provides better performance and determinism compared to the vast majority of general purpose processor-based solutions. FPGAs have applications in various areas of computing such as Signal Processing, TV, Radio, Cellular Telephony, as well as in Computational Neurosciences [23, 24, 32]. They are used in research as a prototyping platform as well as for a final application [34, 36]. The main goal is to develop a complete synchronization method between ANNs using Algebraic Topology, Chaos Theory and Qualitative Theory of ODEs in FPGA reducing computational cost compared to Lyapunov Stability Criterion (LSC) for a possible brain implant that synchronizes signals from neurons. This paper is organized as follows. Section 2 describes the brain and the behavior of neurons defining the concepts of synchronization, chaotic synchronization and how they are present in the processing of the brain. In Sect. 3 the dynamic systems as well as the strange attractors, essential for the understanding of the dynamics of the neural signals and how the synchronization of these systems occurs, are formally presented. In Sect. 4, HNNs will be formally defined, which will model NNNs. In Sect. 5 Homotopies are defined, applications that deform a subset of topological spaces. In Sect. 6 the software and hardware experiment will be presented. In the last Section, the conclusion is presented in addition to future work.

## 2 The Brain

The brain is the main component of the Central Nervous System located inside the skull, divided into more than forty distinct areas, where each one performs a specific activity. We can say that it is the most formidable carbon-based processor. It consists of approximately 86 billion neurons (i.g., nerve cells), where each one has 10,000 connections called synapses [1]. Neurons are divided into sets known as Neural Networks. Each neuron is responsible for processing and disseminating electrical signals, and they control all activities of the organism. They consist of: (a) dendrites, that receive the stimuli being the input terminals; (b) central body, which performs information processing; (c) axons, transmits the stimuli, they are the output terminals.

## 2.1 Synchronization Between Neurons

Synchronization between neurons was observed in the works [6, 7]. In a healthy nervous system, the firing of neurons is not performed randomly. Usually, there is a low-frequency rhythm determining its activity. Studies show that synchronization plays a key role in brain functions involving memory and movement. Understanding how synchronizations occur is fundamental to understand the functioning of the brain and, as a corollary we can identify and avoid cerebral origin pathologies [33]. These pathologies are related to abnormal synchronization of neural firing. Hence, the need to construct theoretical models so that such results can be used for the treatment of diseases. In recent years, a new technique has brought interesting results. The implant of small devices between the skull and the surface of the brain. For Parkinson's disease [2] the device electrodes trigger electrical stimuli to the region of the brain that has symptoms of the disease. Neurology attests that chaotic behaviors are found in the brain [5], both at the neurons and global activity levels. So, unique tools are needed to understand the phenomena of the brain that are introduced in the next Section.

## 3 Dynamical Systems

Natural phenomena, even governed by simple equations have chaotic behavior. The presence of variables at first negligible results in generating unpredictable behaviors in the System. The Nonlinear Dynamic Systems, Chaos Theory, is composed of a range of tools derived from areas of Mathematics and their generality allow it to be used in the solution of problems of linear character [8].

At the end of the nineteenth-century, Poincare, in his works about celestial mechanics it was realized that for some differential equations it was not possible to find an understanding of the quantitative methods. Hence, Poincare introduced other unconventional elements of mathematics to attack such problems as Algebra, Topology and Differential Geometry [9]. Dynamic Systems are divided into two classes. (a) Conservatives: Those who conserve energy, in other words, system energy is constant. In practice, this measure will depend on the nature of the system (e.g., temperature, pressure, volume). (b) Dissipative: Those in which there is the loss of energy through dissipative factors (e.g., temperature drop, volume decrease). The biggest problem in Dynamic Systems theory is to understand its asymptotic behavior. Along the interactions, it can be generated sets of complicated geometry and chaotic behavior. They are named Strange Attractors.

## 3.1 Strange Attractors

Arise after a considerable number of interactions in a Dynamic System. Considering the effects produced, such attractors are extremely sensitive to the simplest variations in the initial conditions of their development, as the interactions advance over time. Thus, a certain pattern of disorder will be developed [9, 10]. The asymptotic behavior of a Dynamic System can evolve to: (a) a fixed point. (e.g. $f : [0, 1] \rightarrow [0, 1]$, $f(x) = ax$ with $a < 1$); (b) a periodic attractor. (e.g., the behavior of a pendulum); (c) a strange attractor. (e.g., Henon, Chua, Plikin, [10]).

## 3.2 Synchronization of Dynamic Systems

Occurs when two or more dissipative Dynamic Systems are coupled. Even if their trajectories deviate exponentially, this phenomenon is obtained experimentally and theoretically well studied [11–14]. We have two types of Chaotic Synchronization. (a) Generalized Synchronization, which occurs when the oscillators are distinguished as Master and Slave. (b) Phase Synchronization, that occurs with identical oscillators where the phase difference is limited, but their amplitudes are not correlated. Formally, (a) is defined as a function that varies with the time when applied to the Slave. It then receives the topological characteristics of the Master. In other words: the Slave is defined from the Master.

# 4    ANNs

They are computational dynamical systems inspired by the central nervous system of an animal. They were first presented in 1943 in the studies of McCulloch and Pitts until in 1949 Heeb published the article: 'The Organization of Behavior' [17], where he proposed a law of organization for neurons. It has several applications in computing such as processing and signals, speech recognition (e.g., images, speech) [18, 19], and in recent years are mainly used to solve Convex Optimization problems. ANNs present three types of learning: (a) Supervised, (b) Not Supervised, and, (c) By Reinforcement. In (a) the training set is presented where outputs will converge to this set. Others network such as (b), update their weights without a training set or any added reinforcement. Nonetheless, (c) has each entry reinforced to adjust the network outputs.

## 4.1    HNNs

First, we formulate the General Equation of the HNN. For this, we consider its synaptic weights $w_{i1}, w_{i2}, \ldots, w_{in}$, which represent the conductances where $n$ is the number of neurons. Consider also $u_1(t), u_2(t), \ldots, u_n(t)$ the inputs (voltages). According to Kirchhoff's Current Law we have:

$$\frac{dv_i(t)}{dt} = -\frac{v_i(t)}{C_i R_i} + \sum_{j=1}^{n} w_{ij} g_i(v_i(t)) + \frac{I_i}{C_i} \tag{1}$$

where $R_i$ is a Leak resistance, $C_i$ a leakage capacitance and $v_i$ is the field induced at the input of the neuron's Activation Function $g_i$ [18].

Later the time delay was introduced $\tau > 0$ in the above equation resulting in our general form:

$$\frac{dv_i(t)}{dt} = \frac{-v_i(t)}{C_i R_i} + \sum_{j=1}^n w_{ij} g_i(v_i(t-\tau)) + \frac{I_i}{C_i} \qquad (2)$$

## 4.2  Synchronization of HNNs

This is a Methodology to develop HNN Synchronization of Chaotic Nature [18]. This Methodology consists of taking an HNN in its general equation given by its matrix form:

$$\phi'(t) = -C\phi(t) + Af(\phi(t)) + Bf(\phi(t-\tau)) + I \qquad (3)$$

By which we will call the Master System. Where $\phi(t)$ is the Neural Network state vector,

$$C = \begin{bmatrix} a_1 & & \\ & \cdots & \\ & & a_n \end{bmatrix} \in R^{n\times n}, a_i > 0,$$

$A, B \in R^{n\times n}$, $f$ is the activation function satisfies the global Lipschitz Condiction i.e.:

$$\|f(x) - f(y)\| \le k\|x - y\| \text{ for all } x, y \in R^n, k \ge 0 \qquad (4)$$

$\tau$ is the delay time, and $I$ is the input vector of the network. We will also take a System with the same Equation corresponding, but with different initial conditions by which we will call Slave System given by:

$$\varphi'(t) = -C\varphi(t) + Af(\varphi(t)) + Bf(\varphi(t-\tau)) + I + U(t) \qquad (5)$$

where $U(t)$ is the portion that over time will generate synchronization [9, 10, 39].

The complete synchronization error dynamics are defined as:

$$e'_t = \phi'(t) - \varphi'(t) \qquad (6)$$

Master and Slave are synchronizable if:

$$\lim_{t\to\infty} \|e_t\| = 0 \qquad (7)$$

In this work, we will synchronize two networks that do not use the $U(t)$ portion. The discovery is an expensive task involving many equations and algebraic formulations, and $U(t)$ depends on the dynamic system [20]. This simplified method will be defined in the next Section, as well as the tool responsible for this simplification.

## 5  Homotopy

Algebraic Topology is an area of Mathematics that has recently presented satisfactory results in applications in Computing. Algebraic Topology presents valuable tools when studying space deformations, symmetries, translations, for example [21].

Among the deformations, we highlight Homotopies. Transformations that deform sub-sets of spaces continuously [22]. The main tool of this work.

Let be $\Omega$ (Omega) a Topological Space [22] and $I = [0, 1]$. Two continuous functions $f$ and $g$ are said Homotopic if there is a continuous function: $H : \Omega \times I \to \Omega$. Such that $H(x, 0) = f(x)$ and $H(x, 1) = g(x)$ for all $x \in \Omega$. The $H$ function is called Homotopy between $f$ and $g$.

In other words, for $t \in [0, 1]$ sufficiently close to 1, $f$ and $g$ are identical.

**Theorem 1:** *Let $\Phi$ and $\Psi$ differentiable dynamic systems synchronizable. There is a Homotopy H such that*: $H(x, 0) = \Phi(x)$ *and* $H(x,1) = \Psi(x)$, $x \in R^n$.

*Proof.* Consider: $k = \frac{1}{t}, t \in [1, \infty)$. Define:

$$H(x, k) = (1 - k)\Phi(x) + k\Psi(x), x \in R^n \tag{8}$$

We have:

$$H(x, 0) = \Phi(x) \text{ and } H(x, 1) = \Psi(x)$$

So:

$$\lim_{t \to \infty} \|\Phi(x) - H(x, k)\| = \|\Phi(x) - \Phi(x)\| = 0 = \lim_{t \to \infty} \|e_t\| \tag{9}$$

Q.E.D.

**Corollary:** *Let $\phi$ and $\psi$ HNNs defined by Eq. (3), synchronizable. There is a Homotopy what synchronizes $\phi$ and $\psi$.*

*Proof.* Trivial.

Theorem 1 guarantees the use of the Eq. (8) in the synchronization implementation justifying the absence of the term $U(t)$ described in Sect. 4. This fact is responsible for the lower complexity of the algorithm since the calculation of $U(t)$ involves many algebraic operations [12–14, 37–39]. The Eq. (8) also ensures that we do not need to have the same dynamic system.

## 6  The Experiment

In this work, as previously mentioned, the method was simplified generating lower computational cost and execution time. Consider the HNN described by Eq. (10):

$$\phi'(t) = -C\phi(t) + Af(\phi(t)) + Bf(\phi(t - \tau)) + I \tag{10}$$

$$\text{Where } C = \begin{bmatrix} 1 & 0 & 0 \\ 0 & 1 & 0 \\ 0 & 0 & 1 \end{bmatrix}, A = \begin{bmatrix} -2 & 0 & 6 \\ -4 & 1 & -1 \\ -6 & -4 & -1 \end{bmatrix},$$

$$f(x) = tanh(x), \quad B = \begin{bmatrix} 1 & 0 & 1 \\ 0 & -3 & 0 \\ 0 & 0 & 0 \end{bmatrix}, I = \begin{bmatrix} 0 \\ 0 \\ 0 \end{bmatrix}$$

The solution of the system of ODEs in (10) has the solution a chaotic attractor Asteriscus described in Fig. 1.

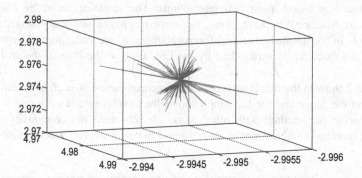

**Fig. 1.** Asteriscus attractor generated in the experiment

The system Master has the following initial conditions: $(-3, 1, 3)$. Take system Slave with the following initial conditions $(4, -1, 5)$.

$$\varphi'(t) = -C\varphi(t) + Af(\varphi(t)) + Bf(\varphi(t - \tau)) + I \tag{11}$$

In the implementation, we used the software MATLAB-2014. For the generation of the Master and the Slave was used the fourth-order Runge-Kutta Method and the synchronization was executed through the Linear Homotopy defined in Eq. (8). Let's take the errors in each coordinate after the synchronization defined by:

$$e_x = x_m - x_s, e_y = y_m - y_s, e_z = z_m - z_s$$
$$\text{where } (x_m, y_m, z_m) \in \text{Master and } (x_s, y_s, z_s) \in \text{Slave.} \tag{12}$$

At this point are observed: (a) the systems to be synchronized need not necessarily to have the same system of equations, but in some cases, the approximation (i.g., synchronization) is only continuous and not differentially continuous; (b) on the other hand, in some cases, the only continuous approximation can be approximated to a differentially continuous. These cases will be exposed in next works.

## 6.1   Implementation

We adopted the number of interactions as a metric for the comparison between our implementation and related works. Since the hardware is different, this presents a straightforward comparison [39].

The experiment was run with an Intel Core 2 Duo processor, 4 GB of memory and Windows 7 Operating System and MATLAB performed the processes of resolution of the ODEs and synchronization in $5.6 \times 10^{-2}$ s. The average time to obtain the complete synchronization occurred in 3 interactions. Being an interaction equal to $1.2 \times 10^{-4}$ s the complete synchronization was obtained in $3.6 \times 10^{-4}$ s, where following [12–14, 37, 38] the systems are synchronized when the average error is less than 0.5 units of measurement.

Considering the errors between the coordinates of system Master and Slave, the graphs below present the performance of the experiment, where the complete synchronization was tested using 500 interactions. The implementation by Homotopy attests to its functionality. Over time, the error reaches a stationary stage, and by Theorem 1, in the implementation of synchronization, we can adopt Homotopy by replacing the calculations established by the LSC used in the state-of-the-art [13, 14, 37–39].

Figure 2 show in the first three graphs the synchronization on each axis between the Master and the Slave and the last graph shows the synchronization error.

Comparing our method with other works, it was used the complexity and the minimum number of interactions of each method to perform a full synchronization. All

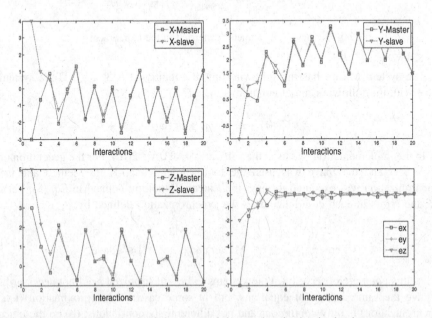

**Fig. 2.** On the three first graphs, it is presented the synchronization in the three axes $x$, $y$, $z$ of Master and Slave respectively till 20 interactions. On the last, we see the errors between the Masters and the Slaves till 20 interactions.

use the Runge-Kutta method for the solution of the equation system. It is defined the total complexity of each method by:

$$TotalComplexity = n(Operations + RK)$$
$$Complexity = n \times Operations$$

where $n$ is the interaction number to the solution (Runge-Kutta). The Homotopy method has a Complexity of $15n$. Table 1 shows the latest results in Chaotic Synchronization and the method involved. Homotopy was applied in each dynamic system of each work. A quick calculation on the complexities of the methods shows that we have a reduced number of operations compared to the others, in addition to a minimum number of interactions for full synchronization (column 'Time' in Table 1) is smaller or equivalent.

**Table 1.** Comparison of results (state-of-the-art)

| Ref | Method | Complexity | Attractor | Time | Our time |
|-----|--------|------------|-----------|------|----------|
| [40] | LSC | $22n$ | Lorenz | 4 | 4 |
| [39] | Adaptive control | $60n + n^2$ | HNN | 5 | 5 |
| [12] | Active disturbance rejection | $50n$ | ODEs | 5 | 4 |
| [13] | LSC* | $(23 + k)n$ ** | Lorenz | 5 | 4 |
| [37] | Probability theory | $kn^2$ ** | N. network | 4 | 4 |

(*) LSC modified. (**) $k >= 1$

## 6.2   Simulation in FPGA

The previous work shows that the FPGA implementation of the synchronizations of dynamic systems generates satisfactory results [25–27], where the synchronization is given by the LSC [26, 28]. In this work, as previously mentioned, the synchronization is simplified, generating a lower computational cost and execution time. Table 2 represents the state-of-the-art on chaotic synchronization problems in FPGAs. The most used attractors and the comparison with our work are presented. Unlike the previous work, the main tool is to approach the problem in general through the Qualitative Theory of ODEs (Algebraic Topology).

**Table 2.** Comparison of results (state-of-the-art)

| Ref | Attractor | FPGA | Qualitative theory |
|-----|-----------|------|--------------------|
| [29] | Lorenz | Yes | No |
| [30] | Chua | Yes | No |
| [31] | Neural network | Yes | No |
| This | General | Yes | Yes |

We chose to implement the chaotic synchronization on an FPGA as a demonstration of the concept. The tests consist of: (a) simple combinatory implementation of the equations using hardware description language (i.e. SystemVerilog); (b) behavioral simulations using ModelSim®; (c) circuit synthesis using QuartusII®; also, (d) programming an Altera® DE4 FPGA board to extract data after the execution.

A simple combinatory implementation was preferred as it can takes advantage of the parallel nature of FPGAs. For this reason, the calculation of each of the three dimensions (i.e. $x$, $y$, $z$) of both Master and Slave networks is concurrent. Subsequently, these outputs are operated together for the chaotic synchronization. An initial validation using behavioral simulation was performed using ModelSim®. After, we developed and compiled a synthesizable version with QuartusII®. Finally, the object code was uploaded to a DE4 board and results extracted. Altera manufactured the board with a Stratix IV chip that has 182,400 Arithmetic LookUp Tables (ALUTs). The circuit uses a total of 10,378 ALUTs, 139 Registers, 30 Digital Signal Processing (DSPs): 24 of 18bit and 6 of 36bit. The circuit has a maximum frequency of operation at 165.62 MHz (as per timing analysis performed by QuartusII). For better stability, the clock frequency was reduced to 150 MHz for the final implementation and results from extraction. At this frequency, the execution time to perform 500 iterations of synchronization was 6.7 microseconds. As the brain operates at a frequency that is an order of magnitude lower [34, 35]. The implementation allows for a real-time application with a brain interface. Nonetheless, it is $8.4 \times 10^3$ times faster than a MATLAB implementation.

Figure 3 shows in the first three graphs the synchronization on each axis. Synchronization errors are displayed in the fourth graph. Note that the error reaches its steady state (minimum value) in the second interaction.

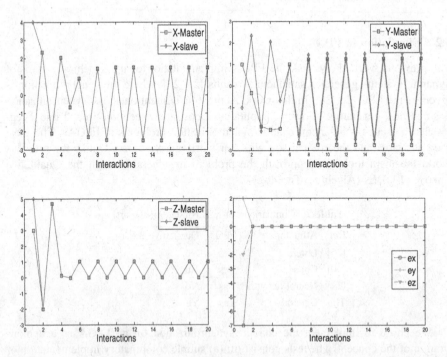

**Fig. 3.** Our implementation using a FPGA. On three first graphs, it is presented the synchronization in the three axes $x$, $y$, $z$ of Master and Slave respectively. The last one present the errors between the Master and the Slave.

### 6.3    Comparison of Results (Hardware vs Software Implentation)

To compare the results of the implementations, software and hardware, due to the change of scale and speed of processing and different metrics were used the Largest Lyapunov Exponent (LLE) of each series $(x, y,$ and $z)$. The Lyapunov Exponents describes the speed at which two near points move away or approach in the course of time subjected to a dynamic system. These values defined in Eq. (13) represent a study in the behavioral sense between the time series, due to their mathematical properties being used for topological comparison among attractors. The LLE is a powerful tool that unites analytical and topological analysis [5, 9, 10].

Let be $\phi : R^n \rightarrow R^n$, differentiable. Define:

$$\lambda(x) = \lim_{n->\infty} \ln \left| \frac{d\phi^n}{dx} \right| \tag{13}$$

$$\text{LLE} = \max\{\lambda(x), \text{for all } x \in R^n\}$$

Table 3 shows the deviation of the LLEs of the implementations, where the maximum deviation is 0.06. So, the implementation in hardware preserves the topological properties of the implementation in software making viable the application in hardware.

**Table 3.** Comparison of results (LLE)

|              | X-Master | Y-Master | Z-Master | X-Slave | Y-Slave | Z-Slave |
|--------------|----------|----------|----------|---------|---------|---------|
| MATLAB (*LEE*) | 1.69 | 1.32 | 1.91 | 1.39 | 1.38 | 1.91 |
| FPGA (*LEE*) | 1.67 | 1.38 | 1.89 | 1.42 | 1.38 | 1.89 |
| $\Delta LEE$ | 0.02 | 0.06 | 0.02 | 0.03 | 0.00 | 0.02 |

## 7    Conclusions

In this paper, it was proposed a methodology to perform the complete synchronization between HNNs simulating NNNs in FPGA using Algebraic Topology tools. The motivation originates in the study of implants in the human brain, with the objective of treating neurological diseases caused by the synchronization deficiencies of the neurons. The methodology is independent of dynamic systems and it has a lower computational cost compared to the state-of-the-art. The application of Homotopy presented satisfactory results, which were verified analytically and experimentally. The results of numerical simulations indicate that the proposed methodology is efficient and convenient for synchronization using FPGAs, where the topological and differential properties of the software implementation were preserved, and the implementation allows for a real-time application with a brain interface.

In future works, we will study simultaneous synchronizations in two or more neural networks. Finally, the experiment confirms the possibility, in theory, that if a brain implant is worked to perform the synchronization of neurons we have appropriate tools for this purpose. To generalize the method for any dynamic system, we first want to analyze systems with high differentiation (second and third derivative) as well as systems defined in larger dimensions. Finally, we want to implement other classes of Homotopy following the implementations in FPGA.

# References

1. Odekerken, V.J., Boel, J.A., Geurtsen, G.J., Schmand, B.A., Dekker, I.P., de Haan, R.J., Schuurman, P.R., de Bie, R.M.: Neuropsychological outcome after deep brain stimulation for Parkinson disease. Neurology **84**, 1355–1361 (2015)
2. Little, S., Pogosyan, A., Neal, S., Zavala, B., Zrinzo, L., Hariz, M., Foltynie, T., Limousine, P., Ashkan, K., Fitzgerald, J., Green, A.L., Aziz, T.Z., Brown, P.: Adaptive deep brain stimulation in advanced Parkinson disease. Ann. Neurol. **74**(3), 449–457 (2013)
3. Bob, P.: Chaos, Cognition and Disordered Brain. Activitas Nervosa Super. **50**(4), 114–117 (2008)
4. Cerutti, S., Carrault, G., Cluitmans, P.J., Kinie, A., Lipping, T., Nikolaidis, N., Pitas, I., Signorini, M.G.: Non-linear algorithms for processing biological signals. Comput. Methods Programs Biomed. **51**(1–2), 51–73 (1996)
5. Maron, G., Barone, D.A.C., Ramos, E.A.: Measuring the differences between spatial intelligence in different individuals using Lyapunov exponents. In: Proceedings of the 7th International Conference on Mass-Data Analysis of Images and Signals, MDA 2012, Berlin (2012)
6. Linas, R.R.: Intrinsic electrical properties of mammalian neurons and CNS function: a historical perspective. Front Cell Neurosci. **8**, 320 (2014)
7. Cabral, J., Luckhoo, H., Woolrich, M., Joensson, M., Mohseni, H., Baker, A., Kringelbach, M.L., Deco, G.: Exploring mechanisms of spontaneous functional connectivity in MEG: how delayed network interactions lead to structured amplitude envelopes of band-pass filtered oscillations. NeuroImage **90**, 423–435 (2014)
8. Frederickson, P., Kaplan, J.L., Yorke, E.D., Yorke, J.A.: The Liapunov dimension of strange attractors. J. Differ. Equ. **49**(2), 185–207 (1983)
9. Viana, M.: Dynamical systems: moving into the next century. In: Engquist, B., Schmid, W. (eds.) Mathematics Unlimited and Beyond. Springer, Heidelberg (2001). https://doi.org/10.1007/978-3-642-56478-9_32
10. Viana, M., Alves, J.F., Bonatti, C.: SRB measures for partially hyperbolic systems whose central direction is mostly expanding. Invent. Math. **140**, 298–351 (2000). Reprinted in the theory of chaotic attractors. Dedicated to J.A. Yorke in commemoration of his 60th birthday. Edited by B.R. Hunt, J.A. Kennedy, T.-Y. Li and H.E. Nusse. Springer Verlag, 443–490 (2004)
11. Pecora, L.M., Carroll, T.L.: Physical review letters. Phys. Rev. Lett. **64**, 821 (1990)
12. Khadra, F.A.: Synchronization of chaotic systems via active disturbance rejection control. Intell. Control Autom. **8**, 86–95 (2017)
13. Ouannas, A., Abdelmaleka, S., Bendoukhaba, S.: Coexistence of some chaos synchronization types in fractional-order differential equations. Electron. J. Differ. Eqn. **2017**(128), 1–15 (2017)
14. Zhang, Q., Lu, J.-A.: Chaos synchronization of a new chaotic system via nonlinear control. Chaos Solitons Fractals **37**(1), 175–179 (2008)
15. González-Miranda, J.M.: Synchronization and Control of Chaos. An Introduction for Scientists and Engineers. Imperial College Press, London (2004)

16. Al-Sawalha, M.M.: Projective reduce order synchronization of fractional order chaotic systems with unknown parameters. J. Nonlinear Sci. **10**, 2103–2114 (2017)
17. Barone, D.A.C.: Sociedades Artificiais: a nova fronteira da inteligência nas máquinas. Bookman, Porto Alegre (2003)
18. Haykin, S.: Redes neurais: princípios e prática. Trad. Paulo Martins Engel. 2. edn. Porto Alegre, Bookman (2001)
19. Hebb, D.O.: Distinctive features of learning in the higher mammal. In: Delafresnaye, J.F. (ed.) Brain Mechanisms and Learning. Oxford University Press, London (1961)
20. Arenas, A., Díaz-Guilera, A., Kurths, J., Moreno, Y., Zhou, C.: Synchronization in complex networks. Phys. Rep. **469**(3), 93–153 (2008)
21. Lima, E.L.: Grupo Fundamental e Espaços de Recobrimento, 4ª edição. IMPA (2012)
22. Lamure, H., Michelucci, D.: Solving geometric constraints by Homotopy. In: Third ACM Symposium on Solid Modeling and its Applications, pp. 263–269. ACM Press (1995)
23. Ahmed, E., Rose, J.: The effect of LUT and cluster size on deep-submicron FPGA performance and density. In: ACM Symposium on FPGAs, FPGA 2000, pp. 3–12 (2000)
24. Lewis, D., Ahmed, E., Baeckler, G., Betz, V., Bourgeault, M., Casshman, D., Galoway, D., Hutton, M., Lane, C., Lee, A., Leventis, P., Marquardt, S., McClintock, C., Padalia, K., Pedersen, B., Powell, G., Ratchev, B., Reddy, S., Sghleicher, J., Stevens, K., Yuan, R., Cliff, R., Rose, J.: The Stratix II logic and routing architecture. In: ACM Symposium on FPGAs, FPGA 2005, pp. 14–20 (2005)
25. Yau, H.T., Pu, Y.C., Li, S.C.: An FPGA-based PID controller design for chaos synchronization by evolutionary programming. Discrete Dyn. Nat. Soc. **2011**, 1–11 (2011)
26. Atoche, A.C., Perales, G.S., Gamboa, A.M., Enseñat, R.A.: Synchronization of chaotic systems: field programable gate array and nonlinear control feedback approach. In: IBERCHIP-2006 (2006)
27. Rajagopal, K., Guessas, L., Vaidyanathan, S., Karthikeyan, A., Srinivasan, A.: Dynamical analysis and FPGA implementation of a novel hyperchaotic system and its synchronization using adaptive sliding mode control and genetically optimized PID control. Math. Prob. Eng. **2017**, Article ID 7307452, 14 p. (2017)
28. Karthikeyan, R., Prasina, A., Babu, R., Raghavendran, S.: FPGA implementation of novel synchronization methodology for a new chaotic system. Indian J. Sci. Technol. **8**, 2 (2015)
29. Vaidyanathan, S., Volos, C.: Advances and Applications in Chaotic Systems. Springer, Berlin (2016). https://doi.org/10.1007/978-3-319-30279-9
30. Muthuswamy, B., Banerjee, S.: A Route to Chaos Using FPGAs: Volume I: Experimental Observations. Springer, Cham (2015). https://doi.org/10.1007/978-3-319-18105-9
31. Park, J., Sung, W.: FPGA based implementation of deep neural networks using on-chip memory only. In: ICASSP 2016 (2016)
32. Cuevas-Arteaga, B., Dominguez-Morales, J.P., Rostro-Gonzalez, H., Espinal, A., Jimenez-Fernandez, A.F., Gomez-Rodriguez, F., Linares-Barranco, A.: A SpiNNaker application: design, implementation and validation of SCPGs. In: Rojas, I., Joya, G., Catala, A. (eds.) IWANN 2017. LNCS, vol. 10305, pp. 548–559. Springer, Cham (2017). https://doi.org/10.1007/978-3-319-59153-7_47
33. WHO: World Health Statistics 2017: Monitoring health for the SDGs. http://www.who.int/gho/publications/world_health_statistics/2017/en/. Accessed 20 June 2017
34. Cassidy, A., Andreou, A.G.: Dynamical digital silicon neurons. In: Biomedical Circuits and Systems Conference, BioCAS 2008, pp. 289–292. IEEE (2008)
35. Ambroise, M., Levi, T., Bornat, Y., Saighi, S.: Biorealistic: spiking neural network on FPGA. In: 2013 47th Annual Conference on Information Sciences and Systems (CISS) (2013)

36. Thomas, D.B., Luk, W.: Biorealistic spiking neural network on FPGA. In: 47th Annual Conference on Information Sciences and Systems, (CISS), pp. 1–6 (2013)
37. Zhu, Q., Song, A., Fei, S., Yang, Y., Cao, Z.: Synchronization control for stochastic neural networks with mixed time-varying delays. Sci. World J. **2014**, Article ID 840185, 10 p. (2014). http://dx.doi.org/10.1155/2014/840185
38. Yue, L., Yixin, Z., Wei, H.: Robust synchronization of uncertain chaotic neural networks with time-varying delay via stochastic sampled-data controller. In: Advanced Information Management, Communicates, Electronic and Automation Control Conference (IMCEC). IEEE (2016)
39. Abdurahman, A., Hu, C., Muhammadhaji, A., Jiang, H.: Adaptive control strategy for projective synchronization of neural networks. In: Cong, F., Leung, A., Wei, Q. (eds.) ISNN 2017. LNCS, vol. 10261, pp. 253–260. Springer, Cham (2017). https://doi.org/10.1007/978-3-319-59072-1_30
40. Park, J.H.: Chaos synchronization of a chaotic system via nonlinear control. Chaos Solitons Fractals **25**, 579–584 (2005)
41. London, M., Roth, A., Beeren, L., Häusser, M., Latham, P.E.: Sensitivity to perturbations in vivo implies high noise and suggests rate coding in cortex. Nature **466**(7302), 123–127 (2010)

# Assessing the Performance of Convolutional Neural Networks on Classifying Disorders in Apple Tree Leaves

Pedro Ballester[⊠], Ulisses B. Correa, Marco Birck, and Ricardo Araujo

Federal University of Pelotas, Pelotas, RS, Brazil
{plballester,ub.correa,mafbirck,ricardo}@inf.ufpel.edu.br

**Abstract.** This paper evaluates the deep learning architecture AlexNet applied to the diagnosis of disorders from leaf images using a recent dataset containing five apple tree disorders. It extends previous work by providing a more extensive testing and a dataset validation by using visualization methods. We show that previous results likely overestimate general accuracy, but that the model is able to learn relevant features from the images.

## 1 Introduction

According to the United Nations, by 2050 the world population is expected to achieve 9 billion individuals while global warming could cut the worldwide crop yields by more than 25% [4]. There are a number of factors that threatens food security besides the climate change, such as the decline in pollinators, pests and pathogens. New technologies and techniques to improve food production must continue to be developed to ensure food security.

Fruits are an important food source to humans and apple is the most consumed fruit after banana. Many disorders can affect the production of apples and the timely diagnosis of such disorders is critical to improve a crop yield. However, a correct diagnosis requires experts which are either not always available or are very costly [3]. Therefore, automating diagnosis could greatly reduce costs and improve apple production.

Many disorders, such as diseases and nutritional deficiencies, affect primarily the leaves of apple trees, making them a natural target for diagnosis, with several previous works tackling this problem from a Machine Learning approach [1,6,8]. In particular, in [3] Convolutional Neural Networks (CNN) are applied to an extensive data set containing five disorders that have a major impact in Brazilian crops: *Glomerella*, Scab, Potassium Deficiency, Magnesium Deficiency and Herbicide Damage. While the results shown were promising, testing was only conducted over a small test set of 75 images. Such small number was justified due to the focus on comparing performance against a set of experts, which were also asked to diagnose the same images; however, such small number can introduce undesired bias, making the reported accuracy (97%) unreliable.

D.A.C. Barone et al. (Eds.): LAWCN 2017, CCIS 720, pp. 31–38, 2017.
https://doi.org/10.1007/978-3-319-71011-2_3

The present work aims at improving the testing methodology used in [3] in order to provide a more robust and reliable measurement of how accurate the proposed approach is. In order to do so, we conduct a more extensive test using far more images and also evaluate learned attributed from the network after training. We show that while accuracy remains high (92%), it is lower than the one reported in the original paper. We also show that the network is indeed able to make use of regions of the images that contain leaf damage, an evidence that the model is learning not only useful, but also correct features.

## 2    Related Works

Several works propose to automatically diagnose plant diseases through image inputs, thus cutting costs of highly specialized personal inspection. Recent works are focused on Convolutional Neural Networks (CNN), a kind of Artificial Neural Network (ANN) capable of learning which features in the input are more significant to the training task.

Despite the increasing use of CNNs, there are still works using classic methods to obtain features from images before applying a machine learning method. In [1] a total of 38 color, shape, and texture features are extracted from leaves pictures. These features are presented as input for a machine learning classifier (Support Vector Machines) obtaining an overall accuracy of 94.22% in their test-set, containing images of leaves affected with powdery mildew, mosaic, and rust.

In [8] authors tested two approaches to image-based diagnosis: a fine-tuned deep network *versus* ad-hoc solution, based on shallow neural network. They had access to a reduced data-set of apple leaves pictures affected by black rot (*Botryosphaeria obtusa*). Botanists labeled dataset images to indicate the disease stage of development. In their experiments, a shallow network was able to achieve accuracy of 79.3% and a VGG16 architecture pre-trained over ImageNet and fine-tuned to the same data set achieved accuracy of 80%.

In [6], a CNN architecture based on CaffeNet and pre-trained on ImageNet is proposed to classify 13 different types of diseases present in leaves of apples, peach, pear, and grapevine trees. This model reaches a reported accuracy of 96.3% when data augmentation techniques is used, such as affine, perspective, and rotation transformations.

In [3], the authors focused on five different disorders present exclusively in apple trees, including not only diseases but also nutritional deficiencies. A dataset with expert labeling was built for the work and made available, which is used in the present work. The authors report a 97% accuracy using an AlexNet architecture pre-trained over ImageNet. However only on a small testing set of 75 images was used. The small number was used due to the need to compare to the classification provided by other experts, which provided evidences that the method could be more accurate than those.

The need for understanding how a neural network makes decisions has been discussed in many works, such as [9]. In particular, identifying which parts of an image influence the network's classification is an ongoing problem in deep learning research. In [10], a heatmap is generated stating which parts of the input participated in the classification process. However, this method gives a broad view, not showing which pixels influenced the most, and tend to not work with Multilayer Perceptron, which AlexNet's output layer is composed of. The approach proposed by [7], on the other hand, allows for the observation of which pixel had the most influence; however the method is not class-discriminative, meaning that if a different label is passed to the method, its result remains the same. In [5], a pixel-specific, class-discriminative approach is proposed that was shown to work well with MLPs. This latter approach is used in the present work.

## 3    Goals and Methodology

Our main goal is to conduct a more extensive analysis of CNNs applied to the problem of classifying disorders in apple trees from leaf images. In particular, we use the same architecture and dataset as the one presented in [3] but conduct a more comprehensive test of its performance and a more thorough analysis of the results. Our specific goals are: (i) to apply a more robust testing methodology to assess the performance of the network and (ii) to better understand image features learned by the network.

In order to accomplish our goals, we use the same AlexNet CNN architecture used in [3]. The network is pre-trained on ImageNet in order to obtain useful filters and then a Multilayer Perceptron is trained at the end of the network to conduct the classification. Adam Optimizer [2] was used with learning rate as $1e - 5$ and a mini-batch size of 5.

While [3] uses only 75 images to test the network, we randomly split the examples into a training set (1000 images) and a test set (500 images), providing a more reliable measurement of generalization capabilities of the model. Both sets are stratified, containing the same number of examples in each class.

We considered two methods to identify and visualize network activations: CAM [10] and Guided Grad CAM [5]. Figure 1 shows an assessment of both methods applied to the dataset. CAM method loses its ability to identify which parts of the image are responsible for a given classification as the information for classifying becomes more spread. Guided Grad CAM, however, more accurately shows which pixels have more influence. Hence, we use this latter method throughout this paper.

Guided Grad CAM is used in two different ways. First, we observe the progression of the visualization against number of epochs. This allows to measure which image parts are becoming more relevant as training progress. Second, we relate regions from the image that are the most important for classifications taking place. We try and relate these regions to spots in the image that are related to the disorder being classified.

| Step | CAM | Guided Grad-CAM |
|------|-----|-----------------|

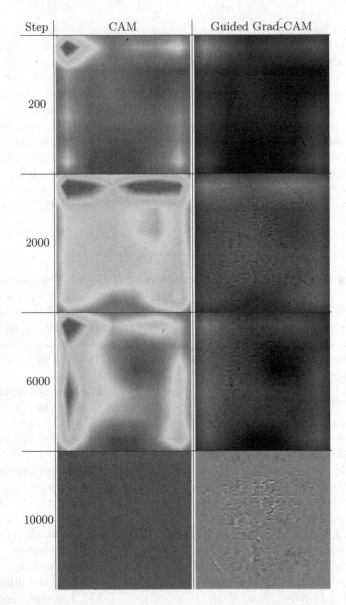

**Fig. 1.** Progression of classic CAM and Guided Grad CAM methods over different number of time steps. Guided Grad CAM is able to better represent the image and active areas.

## 4   Results

Figure 2 shows how accuracy over the test set evolves over the course of training epochs. The network converges after about 2000 epochs and the final accuracy

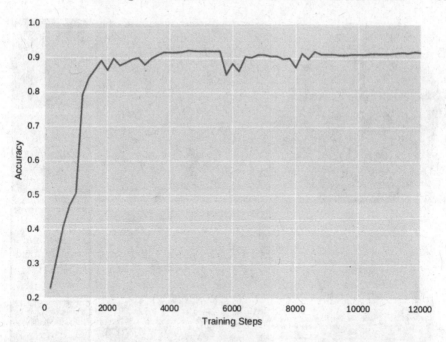

**Fig. 2.** Accuracy over training steps for AlexNet. The results presented correspond to the accuracy over the test set, which do not backpropagate the error. There is no freezing on the convolutional layers during the fine-tuning procedure.

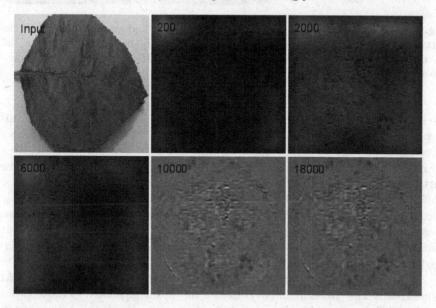

**Fig. 3.** Evolution of Guide Grad CAM during the network's training process. As it becomes better at the classification task, the method shows more clearly which parts of the image are responsible for the correct classification. The numbers represent the number of training steps at each snapshot. (Color figure online)

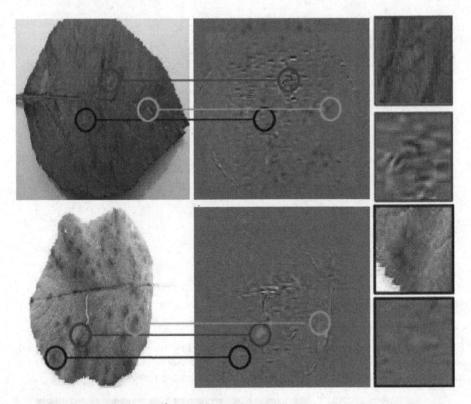

**Fig. 4.** Relations between regions related to the disorder and areas being used by the network. On the right end, a magnification of the red regions of the top image and the blue regions of the bottom image with their Grad-CAM counterparts. (Color figure online)

after 12000 epochs is of 92%, below the reported 97% in [3]. Nonetheless, this is a more accurate estimate of the true performance of the approach since far more examples were used to conduct the test.

Figure 3 depicts the Guided Grad CAM method over a single image. The parts of the image used for classifying a specific class are highlighted by colors, while the unused ones tend to black. In this context, the image shows how the portions of the image used for the networks' decision making process increase over training steps. The observed increase in colored areas resembles how plant disorders affect the leaf, spreading through the surface and creating highly damaged spots.

Figure 4 shows pairs of affected leaves images and their Guided Grad CAM counterparts. The circles highlight regions from the image related to the disorder that had a high response in the network. In the top one, the disorder damage can be identified on the right image by the presence of more colored areas, mostly containing yellow color. The bottom image presents green patterns for most of the regions of interest. This kind of patterns persist through other classes and images, maintaining high response areas in regions directly related the most damaged parts of the leaf.

The above results are evidence that the model is able to learn useful features that correlate well with relevant areas of the image; in contrast, a bias would manifest itself by highlights in areas that are not related to the disorder, such as the image background or mostly in healthy parts of the leaf, none of which is observed.

## 5   Conclusions

This paper provided a more in-depth analysis of the results presented in [3], which was limited due to the need to compare performance against a data set labeled by a panel of experts. In our approach, we aimed at obtaining a more reliable indicator of the performance of the trained model and obtain evidences that this performance is not due to extraneous artifacts in the images.

Our methodology proposed the use of the same data set and architecture used in [3], however using a much larger number (500) of images for testing purposes and a technique, Guided Grad CAM, to allow the visualization of regions of images that contribute to the model's classifications.

Our results allows the conclusion that the AlexNet architecture, pre-trained over ImageNet and fine-tuned to the apple leaves' dataset, is indeed able to classify with a high accuracy the five disorders in the dataset. However, we found that the original accuracy (97%) over the small sample is likely an overestimate of the model's performance, with a 92% accuracy being a more reliable estimator.

Furthermore, by visualizing areas of images that contribute to the classifications, using the Guided Grad CAM technique, we were able to show evidences that the model is learning relevant features from the images, namely damaged leaf areas. No evidences of learned artifacts or bias towards irrelevant image areas (e.g. background) were found.

Finally, we point out directions for further improvements. A cross-validation over the dataset would improve the estimate, even if very costly due to the long training times. However, we believe that it is necessary to test the trained model over a dataset with more diverse examples that is closer to a real-world application - for instance, different lighting conditions, diverse backgrounds and leaf positions.

**Acknowledgements.** This work is supported by CNPq through grant number 407780/2016-5. We gratefully acknowledge the support of NVIDIA Corporation with the donation of the Titan X GPU used for this research.

## References

1. Chuanlei, Z., Shanwen, Z., Jucheng, Y., Yancui, S., Jia, C.: Apple leaf disease identification using genetic algorithm and correlation based feature selection method. Int. J. Agric. Biol. Eng. **10**(2), 74–83 (2017)

2. Kingma, D., Ba, J.: Adam: a method for stochastic optimization. arXiv preprint arXiv:1412.6980 (2014)
3. Nachtigall, L.G., Araujo, R.M., Nachtigall, G.R.: Classification of apple tree disorders using convolutional neural networks. In: 2016 IEEE 28th International Conference on Tools with Artificial Intelligence (ICTAI), pp. 472–476. IEEE (2016)
4. United Nations: Food, July 2017. http://www.un.org/en/sections/issues-depth/food/
5. Selvaraju, R.R., Das, A., Vedantam, R., Cogswell, M., Parikh, D., Batra, D.: Gradcam: why did you say that? Visual explanations from deep networks via gradient-based localization. arXiv preprint arXiv:1610.02391 (2016)
6. Sladojevic, S., Arsenovic, M., Anderla, A., Culibrk, D., Stefanovic, D.: Deep neural networks based recognition of plant diseases by leaf image classification. Comput. Intell. Neurosci. (2016)
7. Springenberg, J.T., Dosovitskiy, A., Brox, T., Riedmiller, M.: Striving for simplicity: the all convolutional net. arXiv preprint arXiv:1412.6806 (2014)
8. Wang, G., Sun, Y., Wang, J.: Automatic image-based plant disease severity estimation using deep learning. Comput. Intell. Neurosci. (2017)
9. Yosinski, J., Clune, J., Nguyen, A., Fuchs, T., Lipson, H.: Understanding neural networks through deep visualization. arXiv preprint arXiv:1506.06579 (2015)
10. Zhou, B., Khosla, A., Lapedriza, A., Oliva, A., Torralba, A.: Learning deep features for discriminative localization. In: CVPR (2016)

# Towards Graffiti Classification in Weakly Labeled Images Using Convolutional Neural Networks

Glauco R. Munsberg[✉], Pedro Ballester, Marco F. Birck, Ulisses B. Correa, Virginia O. Andersson, and Ricardo M. Araujo

PPGC - Federal University of Pelotas (UFPel), Pelotas, RS, Brazil
{grmdsantos,plballester,mafbirck,ub.correa,vandersson, ricardo}@inf.ufpel.edu.br

**Abstract.** Graffiti is an urban phenomenon that can be useful as an indicator of social and economic factors of a geographic region or community. Automatically identifying this urban writings can be useful for understanding cities and their communities. In this paper we investigate the use of Convolutional Neural Networks aiming at classifying weakly labeled images to identify the presence or absence of graffiti art in images. We propose the use of a VGG-16 architecture pre-trained on the ImageNet dataset and show a novel approach to fine-tuning the network over graffiti examples extracted from Flickr. Experiments using this approach show accuracy comparable to that of ImageNet classes.

**Keywords:** Graffiti · Street art · Convolutional Neural Networks · Deep learning

## 1 Introduction

Graffiti is an urban phenomenon that manifest itself by interventions on surfaces like buildings, walls, monuments, and road signs in cities. The nature of graffiti itself is a collection of techniques to change a surface [9], and its art usually contains a high level of abstraction. These urban representations, in general, have some social, politic, or ethnic intrinsic message.

Graffiti can be considered a good indicator of social and economic factors of a geographic region or a community and can be useful to map and track statistics to gather insights of regions of interest. Examples of applications that use graffiti as social indicators are presented in [13,20], where the authors mapped gang and moniker presence in communities using graffiti and similar urban writings. Despite being relatively easy for human beings to detect and classify graffiti in images, it's not a trivial task to automatic systems, due to the high amount of noise and the nature of the surfaces where graffiti are usually projected.

Considering that automatic graffiti classification is an open problem, we propose to investigate computer vision techniques to classify weakly labeled [8] images to identify the presence or absence of graffiti art in images. Recent

© Springer International Publishing AG 2017
D.A.C. Barone et al. (Eds.): LAWCN 2017, CCIS 720, pp. 39–48, 2017.
https://doi.org/10.1007/978-3-319-71011-2_4

advances in machine learning have shown that Convolutional Neural Networks (CNNs) can achieve state-of-the-art results in a variety of applications such as Medical Image Analysis [18], malware detection [22] and object classification in 3D dimension [17].

A large portion of CNNs applied to image recognition tasks make use of architectures that are pre-trained on the popular ImageNet dataset [3]. The reason is that often the target task contains a somewhat small set of images, or with low diversity, to be used for training. By pre-training the network on a large and diverse dataset such as ImageNet, the network is able to acquire kernels that are generally useful for image recognition tasks. The ImageNet dataset contains 1000 classes and 1.2 million images. Then, one can adapt the architecture to a new task by removing the last layer (Fully Connected, FC layer) of the network, which outputs 1000 classes, and adding and training a new layer with the desired number of outputs.

While we follow the general approach and use a network pre-trained on the ImageNet dataset, we propose a different approach to adapt the network to the new task. Instead of removing the FC layer, we instead test the network as-is over a data set of our target task (graffiti identification). We investigate which classes activate the most and then perform a network fine-tuning (i.e. resume training using examples from the new task) but re-associating the most activated class to the class graffiti and the least activated class to the class non-graffiti.

The reasoning to do this is two-fold. First, discarding the last layer also discards information that can be potentially useful for the new task; by keeping it, one allows for the last layer's knowledge to be retained and adapted to the new task, instead of starting from scratch. Second, the method can be easier to be applied as long as the number of classes in the new domain is fewer than the ImageNet domain.

Initial results show that the proposed approach provides faster convergence than replacing the last layer of the network with a binary classifier, achieving accuracies comparable to those found in ImageNet.

This paper is organized as follows. Section 2 presents related works involving graffiti analysis and classification. Section 3 presents our methodology for image acquisition and the proposed model. Section 4 describes the achieved results. Finally, Sect. 5 presents our conclusions so far and possible future works.

## 2   Related Works

There have been several works focused on various levels and implications of graffiti phenomena in society, either by sociological or anthropological sciences [5,6,11] or from computational sciences perspective [7,13,14,20]. On a computational science perspective, the related works focus on development of tools to extract useful information from graffiti and its presence in urban scenarios.

In [7,20] it is presented a graffiti searcher, where given a query image of a graffiti, the $20 * N$ most similar images in a database are returned, aiming to help the law enforcement to identify the graffiti's possible origins. The proposed system makes use of a bag of SIFT (Scale Invariant Feature Transform) features [2] for each image at the database. At each new query, with the image assumed as a graffiti, the bag of features is evaluated and its similarity is measured against other bag of features extracted from the images dataset, returning the most similar ones. Furthermore they provide a list of gang and monikers most associate with these images.

In [13,14], the authors used color based segmentation to obtain text-like graffitis present in images, SIFT descriptors are also used to query a graffiti dataset to detect possible gang connections.

In computer vision feature extraction state-of-the-art techniques, CNNs had their first attempt in [10], but just recently became feasible to apply, achieving superior results compared to the classic techniques. These results, combined with others factors e.g. (computational power, data, etc.) allowed CNNs to be improved and used to solve different problems. They are the first choice as feature extractors when dealing with problems which involves images [4,12,21], and can be fairly used with text [23] or voice [1].

The *ImageNet* dataset [3] is a result of the Large Scale Visual Recognition Challenge 2012 (ILSVRC2012) challenge [15]. The ImageNet dataset is one of the most used benchmarks for image classification and object detection, containing 1000 different classes, organized using the WordNet hierarchy[1]. It contains an average of 1,200 images per class. Due to its rich representativity, it is often used for pre-training CNNs before applying them to new tasks.

## 3  Goals and Methodology

Our main goal is to evaluate how a CNN pre-trained on a generic image data set can be adapted to classify graffiti in images. Our specific goals are (i) to measure whether a CNN trained on ImageNet is able to reliably detect graffiti, even if graffiti is not one of the trained classes and (ii) fine-tune a CNN trained on ImageNet to specifically improve performance on this task.

Our approach to (i) consists of applying a CNN trained on ImageNet and applying it to a data set composed of images containing graffiti and similar images without graffiti, tracking which classes activate for each case. If a single class activate consistently when feeding an image containing graffiti to the network, then the network might be used for this task directly without further adjustments. Of course, since ImageNet does not contain a class for graffiti, this is unlikely but some class might be close enough to provide useful results.

For (ii), we choose an ImageNet class and rename it as "graffiti"; then, we proceed to fine-tune the network using new images, including images with and without graffiti, associating graffiti images to the chosen class. We compare this approach to the more standard fine-tuning procedure, where the FC layer is replaced by a new FC layer with binary output.

---

[1] http://image-net.org/challenges/LSVRC/2016/browse-synsets.

**Table 1.** We build two classes of imagens: The first from tag "graffiti" and the second with tag "street". Both classes and images crawled from Flickr.

| Class | Label of images | Number of images |
|---|---|---|
| Graffiti | "Graffiti" | 468,014 |
| Street | "Street" | 351,338 |

### 3.1 Dataset

In order to fine-tune the model for our new task, we built a dataset composed of two classes: graffiti and non-graffiti Table 1. For the non-graffiti class, we chose to use images of city streets, given that graffiti occur almost exclusively in such setting.

In order to do so, we collected $719,352^2$ images from Flickr, using its API[3], split into $368,014$ images labeled as "graffiti" to represent a diversity of graffiti. $351,338$ images labeled as "street" have also been collected to represent street or the non-graffiti.

The images are *weakly* labeled [19], i.e. labels are noisy due to the use of collaborative tagging by Flickr, where users classify images according to their personal views and experiences. Although these images are weakly classified, an overview of the collected images shows the predominance of reasonable labels.

### 3.2 Experimental Setting

We propose the use of the CNN architecture VGG-16 [16], which achieved state-of-the-art results for the 2016 ImageNet challenge. The architecture is a deeper Neural Network than its predecessors, containing multiple convolutions layers. We initialize the network with weights pre-trained on the ImageNet dataset. In order to train our networks, we use the same parameters on all models, namely Stochastic Gradient Descent with learning rate $\alpha = 1e - 4$, batch size of 128 and a maximum of 150 epochs.

Training and evaluation are conducted by randomly partitioning the dataset into a training set containing 25,000 images of each class and a test set containing 7,500 images of each class.

The first step of our methodology is to apply the network trained only on ImageNet on our graffiti dataset, observing which classes from ImageNet domain activate the most (*one-hot* most activated) and the least (*one-hot* least activated) for each image in the graffiti dataset. By doing so, we aim at verifying which ImageNet classes are most related to the target classes, serving as the basis for the fine-tuning process. Therefore, the classes obtained from this observation are used in the following experiments, as depicted in Fig. 1 and further explained below.

---

[2] Dataset available at http://github.com/ufpeldatalab/graphium/.
[3] www.flickr.com/services/api/.

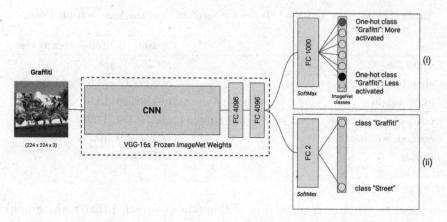

**Fig. 1.** General architecture for our experiments, showing (i) our proposed approach to fine-tuning the network and (ii) the traditional fine-tuning process.

**Pre-trained Remapped Classifier (PRC).** For this approach, we fine-tuned the network maintaining the FC1000 layer as is, remapping the previously found most activated class to the class *graffiti* and the least activated class to the class *street*. For all other classes the output was set to zero. Our goal is to make use of previously obtained knowledge to facilitate learning the new class.

**Pre-trained Inverse Remapped Classifier (PIRC).** In this approach, we reverse the classes *graffiti* and *street* from PRC, mapping graffiti to the least activated ImageNet class. This setting is proposed so as to make it possible to compare against the PRC setting. The reasoning is that by associating to a class that does not strongly activate for our target class, the fine-tuning process will be less efficient in learning the new concept.

**Pre-trained Binary Classifier (PBC).** This is the traditional approach to fine-tuning a network to a different domain. In this setting, the VGG-16's fully connected FC1000 layer designed for the ImageNet domain is replaced by a binary classifier, i.e. a fully connected FC2 layer, for *graffiti* and *street* classes, as depicted in Fig. 1(ii).

## 4    Results

The histograms depicted in Figs. 2 and 3 show the number of times each ImageNet class was the most activated class when images of *graffiti* and *street*, respectively, were presented to the network trained only on the ImageNet dataset.

Table 2 summarizes the results, showing the five most activated classes for *graffiti* and the least activated one. While "freight car" was the class that best

**Table 2.** Five most activated classes for "graffiti" and the least activated one.

| Class | Synset | Top-5 occurrences |
|---|---|---|
| Freight car | n03393912 | 58,955 |
| Comic book | n06596364 | 24,612 |
| Jigsaw puzzle | n03598930 | 13,079 |
| Book jacket, dust cover, dust jacket, dust wrapper | n07248320 | 11,282 |
| Doormat, welcome mat | n03223299 | 10,155 |
| Jacamar | n01843065 | 0 |

responded to *graffiti*, an inspection of the dataset revealed that many graffiti images contained graffiti applied to freight cars and similar transportation vehicles, like subways. Examples are shown in Fig. 4. Therefore, we chose the subsequent class, "Comic book", which does show similarities to graffiti by having strongly saturated colors and well defined borders. The class "Jacamar", a species of bird, was the least activated class.

Table 3 shows the results for the five most activated classes for the *street* images and the least activated one. Classes representing elements commonly present in streets are activated the most, as would be expected, while "Japanese spaniel", a dog breed, was the least activated one.

Nonetheless, we decided to map the class "Jacamar", the least sensitive to graffiti, to the class *street*. This is because *street* is really being used a proxy to the broader class *non-graffiti*, hence we provide an incentive for the network to learn graffiti as the concept instead of "non-street". When using the trained classifier to produce predictions, we consider the activation of any class other than "Comic Book" to represent a classification as a negative example.

Figures 5 and 6 shows loss and accuracy over the training set during the training stage for the three approaches. It is possible to observe that using PRC or PIRC converges quickly (they are very similar and become superimposed on

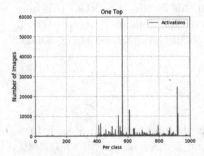

**Fig. 2.** Histogram for the Top-1 "graffiti" activations.

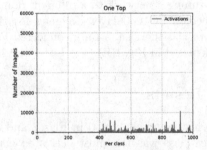

**Fig. 3.** Histogram for the Top-1 "street" activations.

**Table 3.** Five most activated classes for "street" and the least activated one.

| Class | Synset | Top-1 occurrences |
|---|---|---|
| Traffic light, traffic signal, stoplight | n06874185 | 10,533 |
| Cab, hack, taxi, taxicab | n02930766 | 5,909 |
| Street sign | n06794110 | 5,860 |
| Cinema, movie theater, movie theatre, movie house | n03032252 | 5,786 |
| Unicycle, monocycle | n04509417 | 4,971 |
| Japanese spaniel | n02085782 | 0 |

(i)                                (ii)                               (iii)                              (iv)

**Comic Book**                                    **Graffitis**                    **Freight Car with Graffiti**

**Fig. 4.** On the left, example of the class "Comic book" (i), showing saturated color palette and well-defined edges similar to graffiti (ii) (iii). On the right, example images in the dataset, including a freight car containing graffiti (iv). (Color figure online)

the plot), while PBC fails to converge at all for the tested number of epochs. While we believe that PBC would eventually converge for a larger number of epochs, this is evidence that there are benefits to reusing the pre-trained FC layer even if the number of classes do not match the new domain.

Table 4 shows the results of the final networks applied to the hold-out test set. Applying the network without the fine-tuning yields 6.33% true positive rate for graffiti; while this is better than random chance, which would yield 0.1% over the 1000 ImageNet classes, this is still a poor outcome. Replacing the FC layer

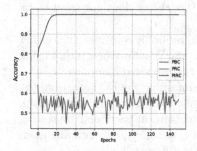

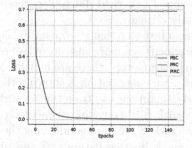

**Fig. 5.** Accuracy during training for each approach.

**Fig. 6.** Loss during training for each approach.

**Table 4.** Top-1 results obtained by applying the specified networks to the hold-out test set. Values shown are for True Positive Rate (TPR) and False Negative Rate (FNR).

| Experiment | TPR | FNR |
|---|---|---|
| ImageNet only | 6.3% | 97.0% |
| PBC | 43.9% | 43.1% |
| PRC | 76.9% | 14.9% |
| PIRC | 72.1% | 12.7% |

with a binary one yields an equally poor outcome since the network fails to converge, yielding results comparable to random choice.

PRC and PIRC shows very similar results. PRC shows a 76.9% true positive rate and 14.9% false negative rate while PIRC shows 72.1% true positive rate and 12.7% false negative rate. This are similar numbers to those observed for ImageNet as a whole [16]. As would be expected from the inversion, PIRC reduces false positives but increases false negatives, but both classes ended up showing very similar results, an evidence that the choice of classes to represent the new domain is not critical.

## 5    Conclusions

In this work we showed early results on the application of Convolutional Neural Networks to the problem of graffiti identification in images. By using a weakly labeled dataset extracted from Flickr and a VGG16 network pre-trained on the ImageNet dataset, we showed that the common approach of replacing the FC layer with one containing the desired number of classes and then training only this layer was unable to converge in the relatively small number of epochs used.

We proposed a more straightforward fine-tuning process where the original FC layer is kept and some of its classes remapped to the new domain. By then fine-tuning the layer, we were able to quickly converge training and obtain reasonable results on a hold-out test set. We showed that the choice of classes to be remapped can yield different outcomes, but this choice does not seem to be extremely critical.

It must be noted that the binary FC layer would probably converge given either enough epochs or a more customized set of parameters. What we show is that using standard parameters, there are benefits to using the proposed approach.

Our main contribution is showing that by maintaining the FC layer from the original domain one can train the network effectively to operate on a new domain. Future work will study this phenomena in more details, in particular by testing against random remapping, allowing the binary FC to converge by running more epochs and adjusting parameters and testing over different domains.

**Acknowledgements.** This work is supported by Google under the Google Research Awards for Latin America, 2016. We gratefully acknowledge the support of NVIDIA Corporation with the donation of the Titan X GPU used for this research.

# References

1. Abdel-Hamid, O., Mohamed, A.R., Jiang, H., Penn, G.: Applying convolutional neural networks concepts to hybrid NN-HMM model for speech recognition. In: 2012 IEEE International Conference on Acoustics, Speech and Signal Processing (ICASSP), pp. 4277–4280. IEEE (2012)
2. Brown, M., Lowe, D.G.: Automatic panoramic image stitching using invariant features. Int. J. Comput. Vis. **74**(1), 59–73 (2007)
3. Deng, J., Dong, W., Socher, R., Li, L.J., Li, K., Fei-Fei, L.: Imagenet: a large-scale hierarchical image database. In: IEEE Conference on Computer Vision and Pattern Recognition, CVPR 2009, pp. 248–255. IEEE (2009)
4. Esteva, A., Kuprel, B., Novoa, R.A., Ko, J., Swetter, S.M., Blau, H.M., Thrun, S.: Dermatologist-level classification of skin cancer with deep neural networks. Nature **542**(7639), 115–118 (2017)
5. Ferrell, J.: Crimes of Style: Urban Graffiti and the Politics of Criminality. Garland, New York (1993)
6. Haworth, B., Bruce, E., Iveson, K.: Spatio-temporal analysis of graffiti occurrence in an inner-city urban environment. Appl. Geogr. **38**, 53–63 (2013)
7. Jain, A.K., Lee, J.E., Jin, R.: Graffiti-ID: matching and retrieval of graffiti images. In: Proceedings of the First ACM Workshop on Multimedia in Forensics, pp. 1–6. ACM (2009)
8. Joulin, A., van der Maaten, L., Jabri, A., Vasilache, N.: Learning visual features from large weakly supervised data. In: Leibe, B., Matas, J., Sebe, N., Welling, M. (eds.) ECCV 2016. LNCS, vol. 9911, pp. 67–84. Springer, Cham (2016). https://doi.org/10.1007/978-3-319-46478-7_5
9. Lassala, G.: Pichação não é pixação: uma introdução à análise de expressões gráficas urbanas. Altamira Editorial, São Paulo (2010)
10. LeCun, Y., Bottou, L., Bengio, Y., Haffner, P.: Gradient-based learning applied to document recognition. Proc. IEEE **86**(11), 2278–2324 (1998)
11. Megler, V., Banis, D., Chang, H.: Spatial analysis of graffiti in san francisco. Appl. Geogr. **54**, 63–73 (2014)
12. Mnih, V., Kavukcuoglu, K., Silver, D., Rusu, A.A., Veness, J., Bellemare, M.G., Graves, A., Riedmiller, M., Fidjeland, A.K., Ostrovski, G., et al.: Human-level control through deep reinforcement learning. Nature **518**(7540), 529–533 (2015)
13. Parra, A., Boutin, M., Delp, E.J.: Location-aware gang graffiti acquisition and browsing on a mobile device. In: Proceedings of the IS&T/SPIE Electronic Imaging on Multimedia on Mobile Devices, p. 830402-1 (2012)
14. Parra, A., Zhao, B., Kim, J., Delp, E.J.: Recognition, segmentation and retrieval of gang graffiti images on a mobile device. In: 2013 IEEE International Conference on Technologies for Homeland Security (HST), pp. 178–183. IEEE (2013)
15. Russakovsky, O., Deng, J., Su, H., Krause, J., Satheesh, S., Ma, S., Huang, Z., Karpathy, A., Khosla, A., Bernstein, M., Berg, A.C., Fei-Fei, L.: ImageNet large scale visual recognition challenge. Int. J. Comput. Vis. **115**(3), 211–252 (2015)
16. Simonyan, K., Zisserman, A.: Very deep convolutional networks for large-scale image recognition. ImageNet Challenge, pp. 1–10 (2014). http://arxiv.org/abs/1409.1556

17. Song, S., Xiao, J.: Deep sliding shapes for amodal 3D object detection in RGB-D images. In: Proceedings of the IEEE Conference on Computer Vision and Pattern Recognition, pp. 808–816 (2016)
18. Tajbakhsh, N., Shin, J.Y., Gurudu, S.R., Hurst, R.T., Kendall, C.B., Gotway, M.B., Liang, J.: Convolutional neural networks for medical image analysis: full training or fine tuning? IEEE Trans. Med. Imaging **35**(5), 1299–1312 (2016)
19. Thomee, B., Shamma, D.A., Friedland, G., Elizalde, B., Ni, K., Poland, D., Borth, D., Li, L.J.: YFCC100M: the new data in multimedia research. Commun. ACM **59**(2), 64–73 (2016)
20. Tong, W., Lee, J.E., Jin, R., Jain, A.K.: Gang and moniker identification by graffiti matching. In: Proceedings of the 3rd International ACM Workshop on Multimedia in Forensics and Intelligence, pp. 1–6. ACM (2011)
21. Vinyals, O., Toshev, A., Bengio, S., Erhan, D.: Show and tell: a neural image caption generator. In: Proceedings of the IEEE Conference on Computer Vision and Pattern Recognition, pp. 3156–3164 (2015)
22. Wang, Q., Guo, W., Zhang, K., Ororbia, I., Alexander, G., Xing, X., Giles, C.L., Liu, X.: Adversary resistant deep neural networks with an application to malware detection. arXiv preprint arXiv:1610.01239 (2016)
23. Zhang, X., Zhao, J., LeCun, Y.: Character-level convolutional networks for text classification. In: Advances in Neural Information Processing Systems, pp. 649–657 (2015)

# Computational Models for the Propagation of Spreading Depression Waves

Guillem Via$^{(\boxtimes)}$, Jean Faber, and Esper Abrão Cavalheiro

Departamento de Neurologia e Neurocirurgia,
Universidade Federal de São Paulo - UNIFESP, São Paulo, Brazil
gviarodriguez@gmail.com

**Abstract.** Spreading Depression (SD) consists on a wave of depressed neural, electrical, activity and near complete depolarization of large neuron populations. It is believed to occur both in compromised and healthy tissue from a broad range of animal species and every structure of the gray matter. Glutamate is long been known to be involved in the ignition of SD. Therefore, despite action potentials are not necessary for the wave propagation, one would expect synaptic processes to play a role in initiating the phenomenon if they are functional. Several detailed and phenomenological computational models have been proposed to simulate the ignition and spread of SD, but few considered synaptic mechanisms. Here we briefly review them, emphasizing macroscopic models that reproduce the wave features and the lack of synaptic transmission. We also propose extensions to a popular model for the wave spread to test whether structural connectivity could aid in stopping the wave and preventing it from engulfing larger portions of the brain.

**Keywords:** Spreading depression · Stroke · Migraine · Reaction-diffusion model · Neural networks

## 1 Introduction

In 1944 Aristides Leão [1,2] noted how direct electrical stimulation could trigger a persistent and spreading depression of the electrical activity in the exposed cortex of rabbits. It was the first time that Spreading Depression (SD) was noticed. Today we know that the depression of electrical activity extends to large populations of neurons and is accompanied by, a near complete neural depolarization, a massive redistribution of ions between the intracellular (ICS) and extracellular space (ECS) of neurons, a very large shift in extracellular electrical potential (of up to $\approx 15\,\mathrm{mV}$), and other physical, chemical and biological alterations [3,4]. The phenomenon has been observed in a broad range of animal species, from vertebrates to invertebrates, and in every gray matter structure from the Central Nervous System. Moreover, it can be triggered by a broad set of chemical, electrical and mechanical stimuli. Despite of its involvement in several disorders like stroke, subarachnoid hemorrhage, trauma injury, and probably migraine,

© Springer International Publishing AG 2017
D.A.C. Barone et al. (Eds.): LAWCN 2017, CCIS 720, pp. 49–60, 2017.
https://doi.org/10.1007/978-3-319-71011-2_5

the mechanisms responsible for it are still far from being well understood. For comprehensive reviews of the pathophysiology and phenomenology of SD see [3,4].

There are four main hypotheses to explain how SD is initiated and how it spreads [3]. In the potassium hypothesis, high extracellular concentrations of this ion increase the excitability of neurons and promote its own further release. Thus, it would trigger a self-regenerative process leading to the large redistribution of ions. In normal conditions, ion pumps at the neuron membrane and glia comprehend a set of buffering mechanisms responsible for clearing these extracellular excesses. However, if concentration exceeds a certain threshold, then the process self-regenerates too fast for the buffering mechanisms to cope with the rises. This reaction-diffusion (RD) process [5] relies both in the diffusion of potassium across the ECS and the reaction it triggers in neighboring tissue, which results in further release of potassium. In favor of this hypothesis, large increases of extracellular potassium concentrations are one of SD defining features [3]. This rise is accompanied by a drastic fall in extracellular concentrations of sodium and chloride, as well as a significant decrease in that of extracellular calcium. In the glutamate hypotheses, it is this substance instead of potassium the one responsible for triggering the above process. The gap-junction and glia hypotheses, consider that the diffusion takes place within the intracellular space of neurons and glia, respectively, across their gap-junctions. There is much evidence supporting the potassium and glutamate hypotheses but none of the four is been ruled out. Finally, the dual hypothesis considers that glutamate or potassium can both ignite the process, each of them doing it in different conditions.

In a seminal work Sugaya et al. [6,7] showed how preventing the generation of action potentials through the application of tetrodotoxin (TTX), a sodium-channel blocker, did not prevent the propagation of SD. Maybe for this reason, since then few computational studies considered the role of synaptic mechanisms on the phenomenon [5]. Only recently Zandt et al. [5] have brought back their possible involvement in the ignition of the SD process, both at the focus of initiation and at neighboring tissue as the wave spreads. Indeed, after the redistribution of ions and the large neural depolarization, synaptic mechanisms cannot be functional, but they can play a role before these changes if they are not inhibited by other means. Different studies support this involvement. It is worth noting that glutamate is been known for long to contribute in the process [3], and blocking synaptic inhibition has been observed to trigger [5] or facilitate [8] SD. Other works involved animal models of migraine. SD is not yet been directly observed during migraine attacks, mainly because of the invasive nature of the techniques allowing for its observation [4]. However, there are several indirect evidences that support the involvement of SD in this disorder. First, computational models have shown how the visual auras observed during migraine attacks are compatible with an SD wave propagating across the visual cortex [9]. These auras consist on alterations of the visual field that spread across it. Second, the BOLD signal observed with fMRI during migraine attacks presents several features characteristic of SD, like an initial increased blood flow (hyperemia) followed by a slight decrease of it with respect to normal values (olygaemia) [4]. Finally, both genetic knockouts and prophylactic treatments in animal models of migraine have shown increases in both (1) resistance to this disorder and (2) threshold for SD initiation, thus drawing a link between the two. Interestingly, two of the most common genetic migraine types, FHM1 and FHM3, have been associated with alterations in neurotransmission [4,10]. In particular, the genetic mutation that seems responsible for FHM1 yields a gain of function in certain voltage-gated calcium channels at excitatory neurons, which results in increased neural excitability [11]. The mutation behind FHM3, instead, is responsible for alterations in the response of voltage-gated sodium channels at inhibitory interneurons, but it is not clear whether these alterations result in a gain or loss of function at these channels [10]. Recently a new computational study showed how, counter-intuitively, increased excitability of inhibitory interneurons can facilitate SD [12]. Altogether, these evidences point towards a direct involvement of neurotransmission in SD.

Here we briefly review the main computational models in the literature used for simulating Spreading Depression, including detailed models in Sect. 2.1 and more abstract ones in Sect. 2.2. In Sect. 3 we describe a popular phenomenological model used for simulating the macroscopic propagating wave of high extracellular potassium, and in Sect. 4 the dynamics it gives rise to. At the end of the section we propose some extensions to the model to capture the transience and decay of the wave. Finally, we dedicate Sect. 5 to draw few conclusions on interesting future directions and open issues for the use of computational models to get a better understanding of the phenomenon.

## 2   Computational Modeling in Spreading Depression

As discussed in the previous section SD involves the interplay of many neural and glial mechanisms. Then, its detailed analysis requires complex models that account for several of them. At the same time, it spreads over distances of up to tenths of centimeters at speeds of few mm/min. It spans, thus, many different time and space scales, which makes it hard to develop a single model capable to account for all of the involved mechanisms. This is why computational and mathematical models of different levels of detail and parameter scales have been developed.

The models can be classified into detailed biophysical and phenomenological ones, according to their degree of complexity, and into macro- and microscopic models, according to the considered space scales [5]. Biophysically realistic models make it easier to establish direct links between the modeled parameters and variables, on the one side, and the real magnitudes and mechanisms, on the other. Bifurcation analysis of these models is also very useful to determine the critical parameter values where transitions occur, e.g. between the healthy resting state and the pathological SD one. Phenomenological models tend to be more abstract and use effective quantities that may not share a direct link with their biophysical counterparts. Their behavior is, though, easier to interpret due to their lower degree of complexity. They also allow for less expensive computational and mathematical treatments. Maybe the most effective way of solving, at least partially, the limitations of these two model types is by linking them. In particular, it is very useful to derive phenomenological models from abstracting more detailed ones, since one can keep trace of the link between derived abstract variables and the underlying biophysical quantities. Here we will briefly summarize some relevant models, for detailed reviews see [5, 7]. We are mainly interested in studying the macroscopic properties of the SD wave, and will thus focus on macroscopic models, which tend to be more phenomenological. However, we start by introducing a few microscopic and detailed ones.

## 2.1 Detailed Models

The first biophysical model used for the study of SD was developed by Tuckwell and Miura [13], which simulated the process at macroscopic scales. It modeled the neural tissue as a continuum of two superposed spaces, representing the ICS and ECS, respectively [7]. The two spaces exchanged a number of ions by means of membrane currents, and the evolution of ion concentrations was described by a set of differential equations, one for each ion specie and space. The model initially considered only potassium and calcium ions in one dimension, but has been extended to account for other ion species in two and three dimensions, and included the glial intracellular space. Detailed microscopic models considering one or few neurons were introduced later which also included only the neuron ECS and ICS [5].

Shapiro [14] presented the most popular model supporting the gap-junction hypothesis. In his model an SD wave could propagate without the need of an ECS across the gap junctions connecting the ICSs of neurons. It is very difficult to interpret due to its high degree of detail and number of variables. For this reason, its results could not be compared to those from other models where potassium propagated across the ECS.

The popular model by Hodgkin and Huxley [15] already proved capable of reproducing well several features of rat pyramidal cell behavior [16]. This is remarkable given that it was developed to simulate a much simpler system, i.e. axons from giant squids. Later multi-compartment models that used the Hodgkin-Huxley approach allowed for a more precise quantitative reproduction of the observed phenomena [5]. Somjen et al. [3,17] considered a detailed multi-compartment model of a single pyramidal cell coupled to a restricted interstitial space in order to study the mechanisms responsible for triggering SD. The model could reproduce realistic tonic firing under electrical stimulation to the soma. They found how disruptions in the potassium buffering mechanisms (i.e. membrane ion pumps and/or glia) prevented the tonic firing from stopping after removing the stimulus. Activation of NMDA-receptor-mediated and persistent sodium currents brought an SD-like depolarization, supporting the importance of these currents in the process. Actually, they showed how one of them alone was enough to trigger the process, but when both acted synergistically the threshold for initiation was lowered and the amplitude and duration of the depolarization were higher. Even more interestingly, the currents could be substituted by any other inward, slowly inactivating, potassium- and/or voltage-gated current and SD would still be triggered. For this, it was also necessary that the inward current indirectly triggered further potassium release to the extracellular space, and that buffering mechanisms could not cope with these releases.

A new model has been recently presented [12] for the simulation of SD which included the effects of synaptic transmission. It is a detailed model that considered a single pyramidal neuron coupled to an inhibitory interneuron and to the extracellular space. Effects of glia were included phenomenologically. They showed how increased excitability at inhibitory neurons could trigger SD, supporting a possible role for the gain of function at certain channels from inhibitory interneurons in triggering SD (see Sect. 1).

As reviewed in [7], it is important to recall some modeling works that did not study directly SD (or not only) but analyzed some important mechanisms that might be involved in it. One example of this is glial spatial buffering. Let us recall that glia are a family of cells that lack the capability of generating action potentials but possess a clear membrane separating their intra- and extracellular spaces. Their ICSs are connected through gap-junctions, membrane channels allowing for the direct transport between cell ICSs. They form, thus, a continuum or syncitium across which potassium and other substances can diffuse. This phenomenon is the so called spatial buffering, believed to contribute to the SD spread, and considered in several models [7]. Porous media and lattice gas models, as well as the lattice Boltzmann method, were used to determine the cell volume fraction and tortuosity. As a result of the exchanges of ions and other substances between the ICS of neurons and glia and the ECS, cells swell during SD. This phenomenon is also been included in a number of models and is been suggested to play an essential role in the propagation of SD. One must note that the reduction of extracellular space in these conditions is been observed to be very large [3].

## 2.2   Reaction-Diffusion Models of SD Propagation

The first phenomenological model of SD was introduced by Grafstein [18], who was also the first one to propose a fundamental role for extracellular potassium concentration on the phenomenon [3,5,7]. This model consists on a partial differential RD equation that describes the evolution of extracellular potassium concentration [5], and was derived following a suggestion from Hodgkin and mathematical analysis by Huxley [5,7]. It shared a striking resemblance to the Fitzhugh-Nagumo model [19] obtained from reducing the dimensionality in the popular model from Hodgkin and Huxley [15]. The similarity comes from the similar shapes of the time-dependent membrane potential in single action potentials and the extracellular potassium concentration at SD ignition. One must remember, though, that the involved time- and space-scales differ in several orders of magnitude [7].

This model was extended to include a second variable that would account for different buffering mechanisms, e.g. of glia and/or increased blood flow [5].

This extended model consists of two coupled differential equations, the original PDE for the extracellular potassium concentration, modified to include the buffering rate variable, and an ordinary differential equation (ODE) to simulate the dynamics of this rate. Reggia and Montgomery [9] used this extended model to study the visual hallucinations or auras reported by migraine patients [5]. For this, Reggia and Montgomery also included a third variable representing the electrical activity of the underlying neural network, and analyzed how a Mexican-hat pattern of connectivity (i.e. excitation between pairs of nearby neurons and inhibition between more distant pairs) would shape this activity. The result was an annular wave of high extracellular potassium preceded by a narrow region of elevated activity that resembled the shape of migraine visual auras. We describe the full model in Sect. 3 and the dynamics it gives rise to in Sect. 4.

However, reports from migraine patients seem to indicate that the wave should break shortly after its initiation and propagate as an arch with decreasing angular extension [20]. This pattern would prevent the wave from engulfing larger portions of the brain [5]. For this reason, Dahlem and Isele [20] included a third variable representing global inhibition mechanisms, instead of the electrical activity, that gave rise to the above mentioned patterns. The evolution of this third variable was described by a third partial differential equation of the reaction-diffusion type that was coupled to the system [5]. This model could reproduce the transience of the spreading waves consistent with experimental observations (see discussion in Sect. 4).

Other simplified models that have been used to study SD include a cellular automata model by Reshodko and Bureš [21] and a model previously developed to simulate wavefronts in cardiac muscle by Wiener and Rosenblueth [22]. Revett et al. [23] also developed a more detailed model based on that by Reggia and Montgomery. Recently the first attempt of coupling a microscopic to a macroscopic model for the simulation of SD was presented [24]. They use Hodgkin-Huxley-like equations for the ion dynamics at the microscopic level and the above reaction-diffusion model for the macroscopic wave characteristics. Also recently O'Connell [25] introduced a detailed microscopic model that included synaptic transmission as well as a broad range of ion-channel types.

To the best of our knowledge, only the models introduced by Desroches et al. [12] and O'Connell [25] considered the effects of synaptic transmission on SD. The possible reasons for this are discussed in Sect. 1, altogether with the motivation to take them into account. Recently, Zandt et al. [5] proposed that the global inhibition variable introduced by Dahlem and Isele [20] could represent certain patterns of structural connectivity instead of the increased blood flow proposed initially by the authors. It is with the aim of tackling this question that we extend the model from Reggia and Montgomery [9] in order to account for effects of electrical activity on extracellular potassium concentrations. Through this interaction we investigate whether a Mexican-Hat structural connectivity can give rise to the wave patterns observed by Dahlem and Isele. In the next section we describe our full model.

# 3   Computational Model of Propagating Waves of Spreading Depression

In this section we describe the computational model from Reggia and Montgomery [9] discussed in Sect. 2.2 with some modifications. In next section we discuss the wave dynamics it gives rise to.

A finite, flat, two-dimensional space is considered, representing a small cortical surface area. The RD equation describing the evolution of extracellular potassium concentration $K(t, x, y)$ and the ODE equation for the buffering rate variable $r(t, x, y)$ are given by

$$\frac{\partial K}{\partial t} = D\nabla^2 K + f(K) \qquad \text{and} \qquad \frac{\partial r}{\partial t} = B((K - K_r) - Cr), \qquad (1)$$

respectively, where $t$ represents time, $x$ and $y$ are the space coordinates, $D$ is the effective diffusion constant, accounting for tortuosity, cell swelling, and possibly also electro-diffusion, $\nabla^2$ is the usual Laplacian operator involving second space derivatives of $K$, while $B$ and $C$ are positive constants modulating the dynamics of $r$. In particular, $r$ increases with rate $B$ when $K > K_r$ and decays to zero with rate $BC$ for $K \approx K_r$. The first term on the right-hand side of the first equation in 1 accounts for the diffusion of potassium while the second term in the same equation is a third order polynomial

$$f(K) = A(K - K_r)(K - K_\theta)(K - K_m)(K + 0.1) - rK \qquad (2)$$

with three zeros: a lower one at normal physiological state (with concentration $K_r$), an intermediate one, threshold for SD activation ($K_\theta$), and a higher excited value ($K_m$), which represents the SD state. These three values of $K$ compose the fixed points of the system, being the lower and higher ones stable, and $A < 0$ is the rate at which the system approaches them.

To simulate electrical activity the two-dimensional flat space is tessellated into an hexagonal grid and each cell is assumed to contain a large population of neurons. The population activity $a_i$ from cell $i$ evolves according to equations

$$\frac{da_i}{dt} = I_i(M - a_i) + c_s a_i \text{ and } I_i = b + \sum_{j \in N_i} c_{ij} a_j, \qquad (3)$$

where $M > 0$ is the maximum activation value, $c_s$ a self-inhibition rate and $I_i$ is the net input received by cell $i$. This input is composed of a basal term $b > 0$ and a contribution from each cell $j$ in the set $N_i$ of neighbors to cell $i$. The weights $c_{ij}$ represent the effective connection strength between the postsynaptic population $i$ and the presynaptic one $j$, and are given by $c_{ij} = c_p g(a_i) / \sum_{n \in N_j} g(a_n)$, with coupling constant $c_p > 0$, gains $g(a_i) = q + \gamma a_i + (1 - \gamma)a_i^2$, and $q$ and $\gamma$ positive parameters.

Activity $a_i$ at each cell $i$ is coupled to extracellular potassium $K_i$ and buffering rate $r_i$ at the same cell by means of substituting the second Eq. 3 by

$$I_i = \left(1.03 - K_i + A_M \exp\left(-\frac{(K_i - K_{0M})^2}{\sigma_M}\right)\right)\left(1.0 - \frac{r_i}{0.04}\right)\left(b + \sum_{j \in N_i} c_{ij} a_j\right), \qquad (4)$$

similar to the expression used in [9]. The constants, obtained phenomenologically, are valid for the case with $D = 0.75$, $A = -0.3$, $K_r = 0.03$, $K_\theta = 0.2$, $K_m = 1.0$, $B = 0.0001$, $C = 10.0$, $A_M = 1.0$, $K_{0M} = 0.12$ and $\sigma_M = 0.0005$.

## 4    Formation and Spread of an SD-like Reaction-Diffusion Wave

We implemented the model equations described in the previous section. Here we briefly describe the nucleation and spread of the wave of high extracellular potassium concentration obtained from them. For this we used the explicit Euler method with a finite difference Forward-in-time-centered-in-space approach to approximate the derivatives [26], for which, in the one-dimensional case, the reaction-diffusion equation is stable under the condition $\Delta t \leq \Delta x^2/2D$, with $\Delta x$ and $\Delta t$ the sizes of the uniform binning for the considered space and time intervals, respectively, and $D$ the diffusion coefficient. We use $\Delta t = 3h^2/32$, with $h$ the distance between opposite sides in each of the hexagonal cells of the uniform grid that divides our two-dimensional space.

As described in [9], the shape and amplitude of the wave of high $K$ is independent on the applied stimulus, as long at is it strong enough to trigger it. Here we describe the nucleation and spread of the wave resulting from the application of a high value of $K$, $K_0 \gg K_r$, at the single central cell for a fixed period of time $t_{stim}$.

The dynamics of nucleation and spread are as follows (see Figs. 1 and 2). First, potassium starts to diffuse from the stimulation point towards its neighborhood and to trigger the reaction that further increases it. The region of increased potassium, thus, starts to grow. If the stimulus is stopped too early, $K$ will diffuse and decay, at these flooded regions, and the system will go back to the homogeneous state with $K = K_r$.

However, if $K_0$ and $t_{stim}$ are large enough, $K$ will experience a sudden rise towards $K_m$ close to the stimulation site, as can be seen in Fig. 2. At this point, it does not matter whether the stimulus is removed or not. $K$ will keep growing towards $K_m$ and trigger the same rise at neighboring tissue. The rise of $K$ is very sharp, so that the point of maximum $K$ is reached very close to the outer edge of the wave. Moreover, this rise is accompanied by an increase of $r$, which, in turn, pulls $K$ towards zero. This results in a smooth decrease of $K$ from the point of maximum $K \approx K_m$ towards the center of ignition (see Fig. 2). This decrease in $K$ is accompanied by a smooth growth in $r$ starting from zero at the outer edge. All of this happens concentrically to the point of stimulation, preserving the symmetry around this point.

At a critical point of low enough $K$ and high enough $r$, $K$ will undergo a sudden drop back to zero, leaving a *hole* of low $K$ near the ignition site. If the stimulus was removed the wave grows around the central point with annular shape. The hole will also grow, keeping the cross section of the wave across any radial plane nearly unchanged. The ring will keep growing and propagating with approximately constant speed and shape.

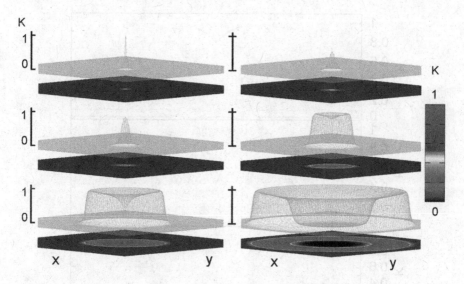

**Fig. 1.** Potassium concentration $K$, normalized to maximum value $K_m$, at the extra-cellular space (ECS) of neurons as a function of space coordinates $x$ and $y$ (without units), across a finite, flat, two-dimensional space representing a small cortical area much larger than the cells themselves. $K$ is initially set very high at a single central point for an extended period of time, starting to diffuse and trigger the release of more potassium into the ECS (top left). After stimulus removal, $K$ initially drops (top right) but soon starts to rise (mid left), since initial $K$ was high enough to trigger the above mentioned release. Buffering rate $r$ (not shown), representing glial buffering mechanisms, starts to rise at the center, gradually reducing $K$ at a growing circular region (mid right and bottom left) until $K$ drops sharply to zero at the center, leaving a hole (bottom right). From this point the ring grows with nearly constant speed and cross-sectional shape. See parameter values at Sect. 3.

Inside the hole $r$ is high and thus $K \approx 0$ even if $K_r > 0$. However, after the wave has passed $r$ will start to decay back to zero and $K$ to recover back to $K_r$. Interestingly, if the stimulus with high $K$ is still present at this point, the elevated $r$ at the central region prevents the nucleation of a second wave. We did not compute long enough times in order to explore whether and how $r$ gradually decays back to zero and allows for a second nucleation.

The coupling between electrical activity and potassium concentration given by Eq. 1 triggers an annular region of very high activity preceding the rapid rise of $K$, while depressed activity is present inside it. Like $K$, within the central hole activity also recovers slowly after the wave passage. Regions of elevated activity always present an annular shape instead of the patches with different shapes shown in [9].

In [20] the authors discuss how this persistent wave that presents no decay can be made unstable. In particular they introduce a third reaction-diffusion equation that makes the system to present a richer repertoire of dynamical states

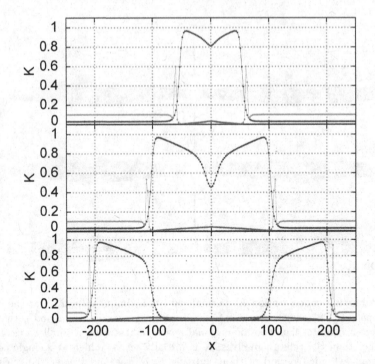

**Fig. 2.** Potassium concentration $K$ (violet) as in Fig. 1 but only across the central line of constant $y$, as a function of position $x$. Plotted are also the buffering rate $r$ (red) and electrical activity $a$ (green). Distributions are computed at times $t = 600, 900$ and $1500$ from top to bottom, which correspond to plots at mid right, bottom left and bottom right from Fig. 1, respectively. (Color figure online)

(see Sect. 2.2). This third equation can also make the annular wave to break along one of its points. After this the open arch decays in angular amplitude as its radius grows, until the two open ends meet and the wave vanishes.

We propose to implement extensions to the above model in order to test the hypotheses proposed in [5]. As discussed in Sect. 2.2 the authors propose that structural connectivity, through electrical activity, could make the annular wave to lose stability and become transient as found in [20]. We plan to couple back the effects of activities $a_i$ on $K$ by adding to Eq. 1 a term dependent on $a_i$ that represents the release of potassium resulting from action potentials and glutamate release.

Further extensions to the model include the substitution of activity Eq. 3 by the more realistic model for neural populations from Renart et al. [27]. This will allow for a more realistic accounting of synaptic mechanisms.

# 5 Concluding Remarks

The interplay of a broad set of mechanisms, interacting at different space and time scales to give rise to the characteristic features of Spreading Depression, requires computational models of different types for a complete comprehension of the phenomenon. Detailed biophysical models are more appropriate for studying the processes responsible for triggering the phenomenon at scales of one or few neurons, while macroscopic phenomenological models are useful for reproducing the characteristics of the macroscopic SD wave.

Here we briefly review some of the most important computational models of SD and propose extensions into an existing one to study the macroscopic characteristics of the wave. These extensions will allow us to study whether certain patterns of structural connectivity can be responsible for providing the SD wave with some of its macroscopic properties. Farther extensions to the model will include the simulation of the neural activity with a mean-field network of leaky-integrate-and-fire neurons.

Previous computational models of spreading depression could reproduce several features of the macroscopic wave of potassium like its all-or-none character, the rapid rise at the front, the elevated electrical activity preceding it, or even its localization and transience. Structural connectivity has been proposed as one of the mechanisms capable of giving rise to this transience, making it to fade away before engulfing large areas of the brain. If true, this would mean that structural connectivity and more in particular inhibition, could play a neuroprotective effect from SD. Extensions to these models will be useful for testing this hypotheses.

**Acknowledgements.** This work was supported by the Instituto de Ciência e Tecnologia (INCT) grant (88887.137596/2017-00) from the INCT call MCTI/CNPq/CAPES/FAPs nr. 16/2014.

# References

1. Leão, A.A.: Spreading depression of activity in the cerebral cortex. J. Neurophysiol. **7**(6), 359–390 (1944)
2. Somjen, G.: Aristides Leao's discovery of cortical spreading depression. J. Neurophysiol. **94**(1), 2–4 (2005)
3. Somjen, G.G.: Mechanisms of spreading depression and hypoxic spreading depression-like depolarization. Physiol. Rev. **81**(3), 1065–1096 (2001)
4. Pietrobon, D., Moskowitz, M.A.: Chaos and commotion in the wake of cortical spreading depression and spreading depolarizations. Nat. Rev. Neurosci. **15**(6), 379–393 (2014)
5. Zandt, B.-J., ten Haken, B., van Putten, M.J., Dahlem, M.A.: How does spreading depression spread? Physiology and modeling. Rev. Neurosci. **26**(2), 183–198 (2015)
6. Sugaya, E., Takato, M., Noda, Y.: Neuronal and glial activity during spreading depression in cerebral cortex of cat. J. Neurophysiol. **38**(4), 822–841 (1975)
7. Miura, R.M., Huang, H., Wylie, J.J.: Cortical spreading depression: an enigma. Eur. Phys. J. Spec. Top. **147**(1), 287–302 (2007)

8. Haglund, M.M., Schwartzkroin, P.A.: Role of NA-K pump potassium regulation and IPSPs in seizures and spreading depression in immature rabbit hippocampal slices. J. Neurophysiol. **63**(2), 225–239 (1990)

9. Reggia, J.A., Montgomery, D.: A computational model of visual hallucinations in migraine. Comput. Biol. Med. **26**(2), 133–141 (1996)

10. Vecchia, D., Pietrobon, D.: Migraine: a disorder of brain excitatory-inhibitory balance? Trends Neurosci. **35**(8), 507–520 (2012)

11. Tottene, A., Conti, R., Fabbro, A., Vecchia, D., Shapovalova, M., Santello, M., van den Maagdenberg, A.M., Ferrari, M.D., Pietrobon, D.: Enhanced excitatory transmission at cortical synapses as the basis for facilitated spreading depression in Ca V 2.1 knockin migraine mice. Neuron **61**(5), 762–773 (2009)

12. Desroches, M., Faugeras, O., Krupa, M., Mantegazza, M.: Modeling Cortical Spreading Depression Induced by the Hyperactivity of Interneurons (2017)

13. Tuckwell, H.C., Miura, R.M.: A mathematical model for spreading cortical depression. Biophys. J. **23**(2), 257–276 (1978)

14. Shapiro, B.E.: An electrophysiological model of gap-junction mediated cortical spreading depression including osmotic volume changes. Ph.D. thesis, University of California, Los Angeles (2000)

15. Hodgkin, A.L., Huxley, A.F.: A quantitative description of membrane current and its application to conduction and excitation in nerve. J. physiol. **117**(4), 500–544 (1952)

16. Zandt, B.-J., Stigen, T., ten Haken, B., Netoff, T., van Putten, M.J.: Single neuron dynamics during experimentally induced anoxic depolarization. J. Neurophysiol. **110**(7), 1469–1475 (2013)

17. Somjen, G., Müller, M.: Potassium-induced enhancement of persistent inward current in hippocampal neurons in isolation and in tissue slices. Brain Res. **885**(1), 102–110 (2000)

18. Grafstein, B.: Mechanism of spreading cortical depression. J. Neurophysiol. **19**(2), 154–171 (1956)

19. FitzHugh, R.: Impulses and physiological states in theoretical models of nerve membrane. Biophys. J. **1**(6), 445–466 (1961)

20. Dahlem, M.A., Isele, T.M.: Transient localized wave patterns and their application to migraine. J. Math. Neurosci. **3**(1), 1 (2013)

21. Reshodko, L., Bureš, J.: Computer simulation of reverberating spreading depression in a network of cell automata. Biol. Cybern. **18**(3), 181–189 (1975)

22. Wiener, N., Rosenblueth, A.: The propagation of impulses in cardial muscle. Arch. Inst. Cardiol. Mex. **16**, 3–4 (1946)

23. Revett, K., Ruppin, E., Goodall, S., Reggia, J.A.: Spreading depression in focal ischemia: a computational study. J. Cereb. Blood Flow Metab. **18**(9), 998–1007 (1998)

24. Gerardo-Giorda, L., Kroos, J.M.: A computational multiscale model of cortical spreading depression propagation. Comput. Math. Appl. **74**(5), 1076–1090 (2017)

25. O'Connell, R.A.: A computational study of cortical spreading depression. Ph.D. thesis, University of Minnesota (2016)

26. Causon, D., Mingham, C.: Introductory Finite Difference Methods for PDEs. Bookboon, London (2010)

27. Renart, A., Brunel, N., Wang, X.-J.: Mean-field theory of irregularly spiking neuronal populations and working memory in recurrent cortical networks. In: Feng, J. (ed.) Computational Neuroscience: A Comprehensive Approach, pp. 431–490. CRC Press (2003)

# Artificial Intelligence

# How Artificial Intelligence is Supporting Neuroscience Research: A Discussion About Foundations, Methods and Applications

Rafael T. Gonzalez(✉), Jaime A. Riascos(iD), and Dante A.C. Barone

Institute of Informatics, Federal University of Rio Grande do Sul,
Porto Alegre, RS, Brazil
rthomazigonzalez@gmail.com, jandresrsalas@gmail.com,
barone@inf.ufrgs.br

**Abstract.** The Artificial Intelligence (AI) research field has presented a considerable growth in the last decades, helping researchers to explore new possibilities into their works. Neuroscience's studies are characterized for recording high dimensional and complex brain data, making the data analysis computationally expensive and time consuming. Neuroscience takes advantage of AI techniques and the increasing processing power in modern computers, which helped improving the understanding of brain behavior. This paper presents some AI techniques, focusing mainly in Deep Learning (DL), as a powerful tool for data analysis. The foundations and basic concepts of some DL models are presented in order to offer a brief understanding to scientists. Likewise, applications of these models on Neuroscience researches are also presented.

**Keywords:** Neuroscience · Neural Networks · Deep Learning

## 1 Introduction

The quick grow of Artificial Intelligence (AI) and its approaches have had an important incidence into several research fields, such as economics, engineering, medicine, so on. This has helped researchers to overcome the limitations raised when analyzing great amounts of data, making possible to explore new horizons in their areas. Modern experimental methods in neuroscience areas such as brain imaging generate vast amount of high dimensional and complex data whose analysis represents a challenge [1]. Machine Learning (ML) models, a sub-set of models of AI that iteratively learn from data without being explicitly programmed where to look [2], are becoming ever more important for extracting reliable and meaningful relationships and for making accurate predictions. Over the past decade, several ML models has been applied to analysis of neuropsychological data such as Magnetic Resonance Imaging (fMRI), Near-Infrared Spectroscopy (NIRS), Electroencephalography (EEG), brain imaging, electromyography (EMG) [3–5] as well as in a high-level, AI can be used to modelling several brain functions [6, 7]. The most popular amongst these methods is Support Vector Machine (SVM) [8]. Despite its popularity, SVM has been criticized for not performing well on raw data and requiring the expert use of design techniques to

D.A.C. Barone et al. (Eds.): LAWCN 2017, CCIS 720, pp. 63–77, 2017.
https://doi.org/10.1007/978-3-319-71011-2_6

extract the less redundant and more informative features (a step known as "feature selection") [9]. These features, rather than the original data, are then used for classification. Most common AI models are considered to be shallow, i.e. they do not create multiple layers of adaptive features and so they are of limited interest to neuroscientists trying to understand perceptual pathways.

More recently, however, it was discovered unsupervised methods for creating multiple layers of features, one layer at a time, without requiring any labels. These methods has proven to be significantly better at creating useful high-level features. These alternative family of ML methods known as Deep Learning (DL) [10] is gaining considerable attention in the wider scientific community [9, 11, 12]. DL methods are a type of representation-learning methods, which means that they can automatically identify the optimal representation from the raw data without requiring prior feature selection. This is achieved through the use of a hierarchical structure with different levels of complexity, which involves the application of consecutive nonlinear transformations to the raw data. These transformations result in increasingly higher levels of abstraction [9]. Inspired by how the human brain processes information, the building blocks of DL neural networks — known as "artificial neurons" — are loosely modelled after biological neurons. Learning is achieved through an iterative process of adjustment of the interconnections between the artificial neurons within the network, much like in the human brain [10]. An essential aspect of DL that differentiates it from other machine learning methods is that the features are not manually engineered; instead, they are learned from the data, resulting in a more objective and less bias-prone process. Besides, the ability to achieve higher orders of abstraction and complexity relative to other ML methods such as SVM makes DL better suited for detecting complex, scattered and subtle patterns in the data [13]. Since high-level features can be more robust against noise in the input data, deep architectures may be more suitable for studying this kind of data than conventional ML methods.

Given the increasingly interest in DL within the field of neuroscience, this review aims to give a brief overview of the foundation of some DL methods and some applications that had been carried out with them in Neuroscience area. In the first part of this review, it is presented the underlying concepts of DL, i.e. Neural Networks. This will be followed by a description of Deep Learning foundations. To achieve this, it will be first presented the concept of Unsupervised Learning, including the common used method called Autoencoder. Afterwards, three Supervised Learning models are presented: Stacked Autoencoder, Convolutional Neural Networks and Recurrent Neural Networks. For each of these models it is presented some of its applications in neuroscience. Finally, the paper is concluded with some future directions.

## 2   Neural Networks

It is estimated that human brain has about 100 billion neurons (nerve cells), connected by an estimated 100 trillion synapses [14]. Neurons share many characteristics with the other cells in the body, but they have unique capabilities for receiving, processing, and transmitting electrochemical signals over the neural pathways that make up the brain's communication system [15]. Given this amazing number of neurons and synapses, it is considered that the human brain operates as a complex, non-linear and parallel computer [16]. An Artificial Neural Network (ANN) is a computational model biologically inspired in the information-processing structures of the human brain. ANNs consist of processing elements called neurons or perceptron and connections between them. These connections are bounded to coefficients (weights) which represent the "memory" of the system. Given the main role of the connections in Neural Networks, they are called connectionist models. Even though ANNs have similarities to the human brain, they are not meant to model it. They are meant to be used for problem-solving and knowledge-engineering in a "humanlike" way [17]. Neural Networks have been applied to many problems of interest to computer science and engineering, such as, pattern classification (the task of assigning an input pattern to one of many prespecified classes), function approximation (finding an estimated value of an unknown function), forecasting (given a set of labeled training patterns in a time sequence, predict the value of a sample at future time) and optimization (find a solution satisfying a set of constraints such that an objective function is maximized or minimized) [18].

Neural Networks are organized, as shown in Fig. 1, in a layer-wise structure where each layer stores increasingly more abstract representations of the data. The first layer $(L_1)$ is the input layer where the data is entered into the model. In neuroimaging, the data can be represented as a one-dimensional vector with each value corresponding to the intensity of one voxel. The last layer $(L_3)$ is the output layer which, in the context of classification, yields the probability of a given subject belonging to one group or the other. The layers between the input and output layers are called hidden layers, with the number of hidden layers representing the depth of the network. Each layer comprises a set of artificial neurons or "nodes" in which each neuron is fully connected to all neurons in the previous layer. Each connection is associated with a weight value $(a_j^i)$, which reflects the strength of each neuron input, much like a synapse between two biological neurons. The structure of neural networks itself allows the transformation of the input space. The consecutive layers perform a cascade of nonlinear transformations that distort the input space allowing the data to become more easily separable.

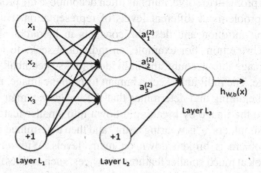

**Fig. 1.** A neural network organized into 3 layers. $L_1$ represents the input layer, $L_2$ is a hidden layer, and $L_3$ is the output layer [19].

Traditionally, neural networks can learn through a gradient descent-based algorithm. The gradient descent algorithm aims to find the values of the network weights that best minimize the error (difference) between the estimated and true outputs. Since Multi-Layer Perceptron (MLP) can have several layers, in order to adjust all the weights along the hidden layers, it is necessary to propagate this error backward (from the output to the input layer). This propagation procedure is called *Backpropagation* [20], and allows the network to estimate how much the weights from the lower layers need to be changed by the gradient descent algorithm. Initially, when a neural network is trained, the weights are set at random. When the training set is presented to the network, this forward propagates the data through the nonlinear transformation along the layers. The estimated output is then compared to the true output, and the error is propagated from the output towards the input, allowing the gradient descent algorithm to adjust the weights as required. The process continues iteratively until the error has reached its minimum value.

Neural networks, which are commonly used in classification tasks, have been applied into a great number of Brain-Computer Interfaces (BCI) studies. In their reviews, Lotte et al. [21], highlighted how NN have been applied to recognize mental states using brain data such as electroencephalogram signals (EEG). It is shown that NN have being used in different problems such as binary or multiclass classification and synchronous or asynchronous BCI. Another application of EEG data and NN classifier is shown in [22]. Bi et al. use NN to translate features extracted from pre-processed and digitized EEG signals into output commands to a robotic system. However, the accuracy of NN classifiers may not be satisfactory in these applications because these models are sensitive to overtraining especially when dealing with such noisy and non-stationary data as EEG [23]. Therefore, careful architecture selection and regularization is required [24].

# 3 Deep Learning

When tasked with a problem to solve, humans often decompose the problem into smaller, easier-to-solve subproblems at different levels of representation. Humans are able to inadvertently exploit intuition and describe concepts in hierarchical ways, based on multiple levels of abstraction. For example, an individual seeks to identity an image. Taking the entire image into account, the individual looks specifically at the important features of the image. The individual sees a human form in the image, notices facial hair, body structure, and clothing and determines that the image is a man. That is, the individual has identified the image by breaking it down into smaller features, such as "has beard", "has broad shoulders", "is wearing a suit" and then determined a classification for the image. The problem is broken down on many levels. Without much conscious thought, humans look at much smaller features of images, such as lines, curves, and edges to determine the higher-level features. These numerous highly-varying, nonlinear features organized into layers are what constitute a deep network [25].

Deep learning generally refers to learning models which use feature hierarchies with many layers. The hidden layers are composed of units that can be used to describe underlying features of the data. In a common facial recognition task, the input layer represents the pixels of the image while the output is the corresponding identity of the face, while the hidden layers can represent low-level features, such as edges and shapes, to high-level features, such as "big eyes" or "short hair". Learning the structure of a deep architecture aims to automatically discover these abstractions, from the lowest to highest levels. Favorable learning algorithms would depend on minimal human effort, while allowing the network to discover these latent variables on its own, rather than requiring a predefined set of all possible abstractions. The ability to achieve this task while requiring little human input is particularly important for higher-level abstractions as humans are often unable to explicitly identify potential underlying factors of the raw input [10]. Thus, the power to automatically learn important underlying features made deep architectures so popular.

At the beginning, training these deep networks using Backpropagation hasn't show good results. The problem stemmed from the fact that as a layer eventually learned a task reasonably well, the learned features were not successfully propagated to successive layers in the network. In these models, the information of the error becomes increasingly smaller as it propagates backward from the output to the input layer, to a point where initial layers do not get useful feedback on how to adjust their weights. This issue was called "the vanishing gradient problem" [26]. In 1992, Hochreiter's mentor, Jürgen Schmidhuber, attempted to solve this problem by organizing a multi-level deep hierarchy which could be effectively pre-trained one level at a time via random initialization and unsupervised learning, followed by a supervised Backpropagation pass for fine-tuning [27]. This method allows each level of the hierarchy to learn a compressed representation of the input observation which is in turn fed into the next level as the successive input. However, while deep architectures were promising, the issue remained that many poor results were suggesting that gradient-based training of randomly initialized supervised deep neural networks easily got stuck in local minima or plateaus [25] and that it becomes increasingly difficult to find a good generalization as the architecture got deeper [28]. In 2006, however, Hinton and colleagues revolutionized the DL field by presenting the idea of "greedy layerwise training" algorithm for construction deep architectures [29]. This method consists of two steps: (1) an unsupervised step, where each layer is trained individually and (2) a supervised step, where the previously trained layers are stacked, one additional layer is added to perform the classification (the output layer), and the whole network parameters are fine-tuned. This breakthrough led to the fast-growing interest in Deep Learning and enabled the development of models that yielded state-of-the-art results in tasks such as handwritten digits classification [29].

## 3.1 Unsupervised Learning

Deep Learning techniques became practically feasible to some extent through the help of Unsupervised Learning (UL). In this context, UL studies how systems can learn to represent particular input patterns in a way that reflects the statistical structure of the overall collection of input patterns. These methods work only with the observed input data, thus there are no explicit target outputs or environmental evaluations associated with each input; rather the unsupervised learner brings to bear prior biases as to what aspects of the structure of the input should be captured in the output [30]. The advantage of learning features from unlabeled data is that, utilizing the plentiful unlabeled data, potentially better features than hand-crafted features can be learned. Both these advantages reduce the need for expertise of the data.

For example, incoming data such as video or speech streams can be encoded in a form that is more convenient for subsequent goal-directed learning. In particular, codes that describe the original data in a less redundant or more compact way can be fed into learning models, whose search spaces may thus become smaller than those necessary for dealing with the raw data. Many methods of UL have been proposed for regularizing NNs, that is, searching for solution computing but simple, low-complexity, which yields higher generalization performance, without overfitting the training data [31].

In [29], Hinton, et al. it is provided an algorithm to pre-train each layer of a deep network using an unsupervised approach. This greedy layer-wise unsupervised learning algorithm first involves training the lower layer of the model with an unsupervised learning algorithm which yields some initial set of parameters for that first layer of the network. That output from the first layer is a reduced representation of the input. This output then acts as the input for the following layer which is similarly trained, resulting in initial parameters for that layer. Again, the output from this second layer is used as the input for the next layer until the parameters for each layer are initialized. The overall output of the network is delivered as the final activation vector. Following this unsupervised pre-training phase of stacked layers, the entire network can then be fine-tuned in the opposite direction using backpropagation in this supervised learning phase.

## 3.2   Autoencoder

Autoencoders are a special case of feedforward networks which comprise of two main components. The first component, i.e. the "encoder", learns to generate a latent representation of the input data, whereas the second component, i.e. the "decoder", learns to use these learned latent representations to reconstruct the input data as close as possible to the original [32]. In its shallow structure, as shown in Fig. 2, an autoencoder

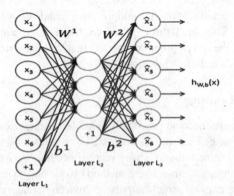

**Fig. 2.** A single layer autoencoder. Layer $L_1$ is the original input data, $L_2$ is the encoded representation of the input data, and $L_3$ is the reconstruction of the input [33].

is comprised of three layers: an input layer ($L_1$), one hidden layer ($L_2$) and an output layer ($L_3$). Moreover, it has a set of parameters ($W$, $b$), where ($W^1, b^1$) represents the weights and biases of the encoder network, and ($W^2, b^2$) represents the weights and biases of the decoder network.

In previous neural networks, labelled data were required to act as training examples essential to the backpropagation fine-tuning pass as those labels were used to readjust the connection weights. However, since an autoencoder does not make use of labels, its training is an unsupervised learning process. An autoencoder neural network performs backpropagation by setting the target output values equal to the input values, and thus it is trained to minimize the discrepancy between the data and its reconstruction. In other words, it is trying to learn an approximation to the identity function, so as to output $\hat{x}$ that is similar to $x$ [33]. While this may seem like a trivial learning task, placing constraints on the network can reveal interesting structure of the data. An example of a constraint is a limitation to the number of hidden units in the hidden layer, thus forcing the network to learn a compressed representation of the input. This method allows for the discovery of internal representations of the data that rely on fewer intermediate features. Another constraint that can be applied to the network could be the sparsity of hidden units that are activated. Sparsity is a useful constraint when the number of hidden units is large (even larger than the number of input values) that can allow for the discovery of interesting structure of the data [32]. A sparse autoencoder has very few neurons that are active. A neuron in an artificial neural network is informally considered "active" if its output value is close to 1, while it is considered "inactive" if its output value is close to 0. The concept of creating a sparse autoencoder involves constraining the most of the neurons to be inactive [32]. As a result, even with many hidden units, the data is constrained, forcing the network to learn the important features of the data in order to reconstruct it.

Hinton et al. defined an autoencoder as a nonlinear generalization of Principal Components Analysis (PCA), which is restricted to linear mapping [34]. If no non-linear function is used in the encoder network of the AE and the number of neurons in the hidden layer is of smaller dimension than that of the input then PCA and AE can yield similar results. On the other hand, in the case of the neurons in the hidden layer is greater than the input size, the AE is transforming the input data from one feature space to another wherein the data in the new feature space disentangles factors of variation. PCA has the advantage that it can work with very little data, while Autoencoders can overfit if not enough data is available.

Autoencoders have being widely applied in studies that use a range of neuroimaging modalities including structural Magnetic Resonance Imaging (sMRI), resting-state functional MRI (rsfMRI) and positron emission tomography (PET) [35, 36]. In [37], Payan and Montana used Sparse Autoencoders and Convolutional Neural Networks to predict the Alzheimer's Disease (AD) status of a patient based on 3D MRI scan of the brain. Their proposed method outperforms several other classifiers reported in the literature and produce state-of-art results. Suk et al. [38] developed an approach which classifies people with Mild Cognitive Impairment (MCI) and healthy controls using a deep autoencoder to extract hierarchical nonlinear relations among brain regions, whilst modelling the inherent functional dynamics of rsfMRI data. This was also one of the few studies in which the same DL model was tested against and surpassed other competing models in two independent datasets, thus providing evidence of replicability, a crucial feature for diagnostic tools.

### 3.3    Stacked Autoencoder

Based on the fact that Autoencoders are automatic features extractors, they can also be stacked to create a deep structure to increase the level of abstraction of learned features. Thus, a Stacked Autoencoder (SAE) is a neural network consisting of multiple layers of Autoencoders [10]. In this case, the network is pre-trained, i.e. each layer is treated as a shallow autoencoder, generating latent representations of the input data. These latent representations are then used as input for the subsequent layers before the full network is fine-tuned using standard supervised learning algorithm [25]. With this deep architecture, learning-feature hierarchies are formed by using lower-level learned features to compose higher levels of the hierarchy. The first layer of a SAE tends to learn first-order features in the raw input (such as edges in an image), the second layer tends to learn second-order features corresponding to patterns in the appearance of first-order features (for example, contour or corner detectors) and, following this logic, higher layers to learn even higher-order features [25].

SAE have been successfully used in studies of brain psychosis diagnostics. In [39], it is used a SAE to extract latent features from neuroimaging data (sMRI, PET and CSF), which were then used to predict clinical data and class labels. The resulting learned features were combined with original low-level features to build a robust model for AD/MCI classification that achieved high diagnostic accuracy. Another research area of high clinical interest is prediction of response to treatment. In several psychiatric and neurological disorders, a better understanding of why some patients benefit from a certain treatment whereas others do not, could help clinicians make more-effective treatment decisions and improve long-term clinical outcomes [40]. In [41], it presented an algorithm that distinguished between patients with temporal lobe epilepsy (TLE) who did and did not benefit from surgical treatment. The proposed uses a SAE to extract meaningful features from diffusion-weighted images (DWI) while a SVM is chosen as the classifier.

### 3.4    Convolutional Neural Networks

Convolutional Neural Networks (CNNs) are very similar to ordinary Neural Networks previously explained. They are a special type of feedforward neural networks that were biologically-inspired by the visual cortex [42]. The visual cortex contains a complex arrangement of cells that are sensitive to small sub-regions of the visual field, called a receptive field. The sub-regions are tiled to cover the entire visual field. These cells act as local filters over the input space and are well-suited to exploit the strong spatially local correlation present in natural images [43].

In addition to the input and output layers, CNN can mainly comprise of three types of layers: a convolutional layer, a pooling layer, and a fully-connected layer [44]. The first one acts as feature identifiers, i.e. they are filters that extract characteristics from input data (such as edges and curves). As the number of convolutional layers increase, more complex features can be represented (such as hands or ears). A common operator used together with convolution is pooling, which combines nearby values in input or feature space through a sample-based discretization process. The objective is to down-sample an input representation, reducing its dimensionality and allowing for assumptions to be made about features contained in the sub-regions binned. This process helps over-fitting by providing an abstracted form of the representation. As well, it reduces the computational cost by reducing the number of parameters to learn. Finally, the fully-connected layers are similar to the hidden layers from the conventional MLP where the neurons are connected to all neurons from the previous layer. The CNN arranges its neurons in three dimensions (width, height, depth), these values are proportional to the size and channels of the input. Every layer in the CNN transforms the 3D input volume to a 3D output volume of neuron activations. Figure 3 exposes a normal Neural Network and a CNN.

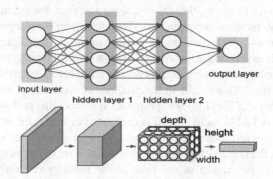

**Fig. 3.** Top: common Neural Network. Bottom: CNN [44]

The CNN's architecture is designed to take advantage of the 2D structure of an input image (or other 2D input such as a speech signal) and to encode certain properties into the architecture. Another benefit of CNNs is that they are easier to train and have many fewer parameters than fully-connected layers networks with the same number of hidden units. Many neurally-inspired models can be found in the literature, such as the NeoCognitron [45], HMAX [46] and LeNet-5 [43]. Many present competition-winning models are based on CNNs. A BP-trained CNN set a new MNIST (handwritten digit recognition dataset) record of 0.39% [47], using training pattern deformations but no unsupervised pre-training.

As was exposed above, CNN was mainly designed for images as input; therefore, the processing, analysis and classification of neuroimaging may turn out to be among the most important applications of Deep Learning, because it could not only save lots of money, but also make expert diagnostics more accessible. For example, an important application of CNNs on cancer diagnosis was presented in [48], the authors successfully use deep max-pooling CNNs to detect mitosis in breast histology images. Likewise, several authors have worked with CNN using either MRI and functional MRI (fMRI) as data input. Sarraf and Tofighi [49] used fMRI data for classification of

Alzheimer's Disease (AD). This work suggests that the shift and scale invariant features extracted by CNN and a deep learning classification are the most powerful method to distinguish clinical data from healthy data in fMRI. Another study performed by Liu et al. [50] use fusing multi-modal neuroimaging features to aid the diagnosis of AD. This framework presented the potential to require less labelled data and therefore a performance gain was achieved in both binary classification and multi-class classification of AD. Moreover, van der Burgh et al. [51] used clinical characteristics in combination with MRI data to predict survival of Amyotrophic lateral sclerosis (ALS) patients using deep learning. This approach reached an accuracy of 84.4%.

### 3.5  Recurrent Neural Networks

Recurrent Neural Networks (RNNs) are obtained from the feedforward network by connecting the neurons' output to their inputs [52]. They are called recurrent due that perform the same task for every element of a sequence; thus, the output of the network is depended on the previous computations. The short-term time-dependency is modelled by the hidden-to-hidden connections without using any time delay-taps. They are usually trained iteratively via a procedure known as backpropagation-through-time (BPTT). RNNs can be seen as very deep networks with shared parameters at each layer when unfolded in time. This results in the problem of vanishing gradients [53]. Long Short-Term Memory (LSTM) [54] was proposed to resolve this problem for Recurrent Neural Networks. A LSTM memory cell is composed of four main elements: an input gate, a neuron with a self-recurrent connection, a forget gate and an output gate. The input gate controls the impact of the input value on the state of the memory cell and the output gate controls the impact of the state of the memory cell on the output. The self-recurrent connection controls the evolution of the state of the memory cell and the forget gate determines how much of prior memory value should be passed into the next time step. Depending on the states of these gates, LSTM can represent long-term or short-term dependency of sequential data.

Recently, LSTM RNNs won several international competitions and set numerous benchmark records. A stack of bidirectional LSTM RNNs broke a famous TIMIT speech (phoneme) recognition record [55]. For optical character recognition (OCR), LSTM RNNs outperformed commercial recognizers of historical data [56]. LSTM-based systems also set benchmark records in language identification [57], medium-vocabulary speech recognition [58], and text-to-speech synthesis [59].

RNN has being widely used for modelling various neurobiological phenomena, considering anatomical, electrophysiological and computational constraints. The computational power of RNNs comes from the fact that its neuron's activity is affected not only by the current stimulus (input) to the network but also by the current state of the network, it means that into the network will keep on traces of past inputs [61]. Thus, the RNNs are ideally suited for computations that unfold over time such as holding items in working memory or accumulating evidence for decision-making. Nevertheless, it is this feature that difficult to train RNNs (as was explained above). For example, Barak [60] presents RNNs as a versatile tool to explain such neural phenomena including several constraints. He exposes how combining trained RNNs with reverse engineering can represent an alternative framework for neuroscience modelling, potentially serving as a powerful hypothesis generation tool. Moreover, Rajan et al. [61] show RNN models of neural sequences of memory based on decision-making tasks generated by minimally structured networks. It suggests that neural sequences activation may provide a dynamic mechanism for short-term memory, which comes from largely unstructured network architectures. In the same way, Güçlü and van Gerven [62] show how RNNs are a well-suited tool for modelling the dynamics of human brain activity. In their approach, they investigated how the internal memories of RNNs can. be used in the prediction of feature-evoked response sequence, which are commonly measured by fMRI. Likewise, Susillo et al. [63] use RNNs to generate muscle activity signals (electromyography, EMG) to explain how the neural responses in motor cortex. They started with the hypotheses that motor cortex reflects a dynamic, which is used for generating temporal commands. Thus, the RNNs are used to transform simple inputs into temporal and spatial complex patterns of muscle activity. Finally, Viera et al. [64] made a deep review and discussion of how deep learning has been used to investigate the neuroimaging correlates of psychiatric and neurological disorders, explaining and comparing different methods and applications.

## 4   Conclusion

This paper offers a discussion about the significant support that AI offers to Neuroscience research. The capacity of DL models to learn complex and abstract representations through nonlinear transformations provides preliminary evidences supporting its potential role in the future development of Neuroscience. Some learning methods and its applications were presented here, mainly focusing on Neural Network and Deep Learning. The models were explained since their foundations until their current state. In summary, Neural Networks have being used in classification, regression and clustering problems, playing an important role in BCI studies. DL models offer alternative and powerful tools for modelling neural activity (RNN), processing neuroimaging (CNN and AE), predicting clinical data and class labels (SAE) and performing dimensionality reduction (AE).

Nevertheless, several improvements will be required before the full potential of DL in Neuroscience can be achieved. Given the complexity of DL models, we need to favor studies that use large data samples. A possible way of achieving this is through multi-center collaborations, in which data is collected using similar criteria and scanning protocols across sites. Moreover, the integration of CNN and RNN is likely to lead to significant advances in DL in the next few years [65]. In neuroimaging, for example, this integration could be particularly useful for analyzing fMRI data, as it would allow the detection of intricate spatial patterns while simultaneously modelling the temporal component of some signals. Finally, it is so important to highlight the necessity of a strong collaboration between Neuroscience and DL. Neuroscientists are interested in DL models as an important tool for understanding the brain behavior and, likewise, computer scientists are taking into account brain theories when creating new models and techniques.

# References

1. Helmstaedter, M.: The mutual inspirations of machine learning and neuroscience. Neuron **86** (1), 25–28 (2015)
2. Bishop, C.M.: Pattern Recognition and Machine Learning. Springer, Heidelberg (2007). ISBN-10: 0387310738, ISBN-13: 978-0387310732
3. Patel, M.J., Khalaf, A., Aizenstein, H.J.: Studying depression using imaging and machine learning methods. NeuroImage: Clin. **10**, 115–123 (2016)
4. Khachab, M., Mokbel, C., Kaakour, S., Saliba, N., Chollet, G.: Brain imaging and machine learning for brain-computer interface. In: Biomedical Imaging, InTech (2010)
5. Lemm, S., Blankertz, B., Dickhaus, T., Müller, K.-R.: Introduction to machine learning for brain imaging. NeuroImage **56**(2), 387–399 (2011)
6. Yamins, D.L.K., DiCarlo, J.J.: Using goal-driven deep learning models to understand sensory cortex. Nat. Neurosci. **19**, 356–365 (2016)
7. Kasabov, N.K.: NeuCube: a spiking neural network architecture for mapping, learning and understanding of spatio-temporal brain data. Neural Netw. **52**, 62–76 (2014)
8. Vapnik, V.: The Nature of Statistical Learning Theory. Springer, New York (1995)
9. LeCun, Y., Bengio, Y., Hinton, G.: Deep learning. Nature **521**, 436–444 (2015)
10. Bengio, Y.: Learning deep architectures for AI. Found. Trends Mach. Learn. **2**, 1–127 (2009)
11. Arbabshirani, M.R., Plis, S., Sui, J., Calhoun, V.D.: Single subject prediction of brain disorders in neuroimaging: promises and pitfalls. Neuroimage **145**, 137–165 (2016)
12. Calhoun, V.D., Sui, J.: Multimodal fusion of brain imaging data: a key to finding the missing link(s) in complex mental illness. Biol. Psychiatry: Cogn. Neurosci. Neuroimaging **1**, 230–244 (2016)
13. Plis, S.M., Hjelm, D.R., Salakhutdinov, R., Allen, E.A., Bockholt, H.J., Long, J.D., Johnson, H.J., Paulsen, J.S., Turner, J., Calhoun, V.D.: Deep learning for neuroimaging: a validation study. Front. Neurosci. **8**, 1–11 (2014)
14. Herculano-Houzel, S.: The remarkable, yet not extraordinary, human brain as a scaled-up primate brain and its associated cost. In: Proceedings of the National Academy of Sciences, USA, vol. 109 (Supp 1), pp. 10661–10668 (2012)
15. Herculano-Houzel, S.: The human brain in numbers: a linearly scaled-up primate brain. Front. Hum. Neurosci. **3**, 31 (2009). https://doi.org/10.3389/neuro.09.031.2009
16. Nygren, K.: Stock prediction - a neural network approach. Master thesis, Royal Institute of Technology, KTH (April 2004)
17. Kasabov, N.K.: Foundations of Neural Networks, Fuzzy Systems, and Knowledge Engineering. MIT Press, Cambridge (1996)
18. Jain, A.K., Mao, J., Mohiuddin, K.M.: Artificial neural networks: a tutorial. Computer **29**(3), 31–44 (1996)
19. Ng, A., Ngiam, J., Foo, C., Mai, Y., Suen, C.: UFLDL Tutorial (2013) Retrieved from Stanford Deep Learning: http://ufldl.stanford.edu/wiki/index.php/Neural_Networks
20. Rumelhart, D.E., Hinton, G.E., Williams, R.J.: Learning representations by back-propagating errors. Nature **323**(6088), 533–536 (1986)
21. Lotte, F., Congedo, M., Lécuyer, A., Lamarche, F., Arnaldi, B.: A review of classification algorithms for EEG-based brain–computer interfaces. J. Neural Eng. **4**(2), R1–R13 (2007)
22. Bi, L., Fan, X.A., Liu, Y.: EEG-based brain-controlled mobile robots: a survey. IEEE Trans. Hum. Mach. Syst. **43**(2), 161–176 (2013)

23. Balakrishnan, D., Puthusserypady, S.: Multilayer perceptrons for the classification of brain computer interface data. In: Proceedings of the IEEE 31st Annual Northeast Bioengineering Conference (2005)
24. Jain, A.K., Duin, R.P.W., Mao, J.: Statistical pattern recognition: a review. IEEE Trans. Pattern Anal. Mach. Intell. 22(1), 4–37 (2000)
25. Bengio, Y., Lamblin, P., Popovici, D., Larochelle, H.: Greedy layer-wise training of deep networks. In: Advances in Neural Information Processing Systems, p. 153 (2007)
26. Hochreiter, S. Untersuchungen zu dynamischen neuronalen Netzen. Diploma thesis. Institut f. Informatik, Technische Univ. Munich (1991)
27. Schmidhuber, J.: Learning complex, extended sequences using the principle of history compression. Neural Comput. 4(2), 234–242 (1992)
28. Larochelle, H., Bengio, Y., Louradour, J., Lamblin, P.: Exploring strategies for training deep neural networks. J. Mach. Learn. Res. 10, 1–40 (2009)
29. Hinton, G.E., Osindero, S., Teh, Y.W.: A fast learning algorithm for deep belief nets. Neural Comput. 18(7), 1527–1554 (2006)
30. Barlow, H.B.: Unsupervised learning. Neural Comput. 1, 295–311 (1989)
31. Baum, E.B., Haussler, D.: What size net gives valid generalization? Neural Comput. 1(1), 151–160 (1989)
32. Hinton, G., Salakhutdinov, R.: Reducing the dimensionality of data with neural networks. Science 313(5786), 504–507 (2006)
33. Ng, A., Ngiam, J., Foo, C., Mai, Y., Suen, C.: UFLDL Tutorial (2013). Retrieved from Stanford Deep Learning: http://ufldl.stanford.edu/wiki/index.php/Autoencoders_and_Sparsity
34. Calhoun, V.D., Silva, R.F., Adali, T., Rachakonda, S.: Comparison of PCA approaches for very large group ICA. Neuroimage 118, 662–666 (2015). https://doi.org/10.1016/j.neuroimage.2015.05.047
35. Liu, S., Liu, S., Cai, W., Che, H., Pujol, S., Kikinis, R., Feng, D., Fulham, M.J.: Multimodal neuroimaging feature learning for multiclass diagnosis of Alzheimer's disease IEEE Trans. Biomed. Eng. 62, 1132–1140 (2015)
36. Han, X., Zhong, Y., He, L., Philip, S.Y., Zhang, L.: The unsupervised hierarchical convolutional sparse auto-encoder for neuroimaging data classification. In: Guo, Y., Friston, K., Aldo, F., Hill, S., Peng, H. (eds.) BIH 2015. LNCS, vol. 9250, pp. 156–166. Springer, Cham (2015). https://doi.org/10.1007/978-3-319-23344-4_16
37. Payan, A., Montana, G.: Predicting Alzheimer's disease: a neuroimaging study with 3D convolutional neural networks. arXiv preprint arXiv:1502.02506 (2015)
38. Suk, H.I., Wee, C.Y., Lee, S.W., Shen, D.: State-space model with deep learning for functional dynamics estimation in resting-state fMRI. Neuroimage 129, 292–307 (2016)
39. Suk, H.I., Shen, D.: Deep learning-based feature representation for AD/MCI classification. In: Mori, K., Sakuma, I., Sato, Y., Barillot, C., Navab, N. (eds.) MICCAI 2013. LNCS, vol. 8150, pp. 583–590. Springer, Heidelberg (2013). https://doi.org/10.1007/978-3-642-40763-5_72
40. Mechelli, A., Prata, D., Kefford, C., Kapur, S.: Predicting clinical response in people at ultra-high risk of psychosis: a systematic and quantitative review. Drug Discov. Today 20, 924–927 (2015)
41. Munsell, B.C., Wee, C.Y., Keller, S.S., Weber, B., Elger, C., da Silva, L.A.T., Nesland, T., Styner, M., Shen, D., Bonilha, L.: Evaluation of machine learning algorithms for treatment outcome prediction in patients with epilepsy based on structural connectome data. Neuroimage 118, 219–230 (2015)
42. LeCun, Y., Bottou, L., Bengio, Y., Haffner, P.: Gradient-based learning applied to document recognition. In: Proceedings of the IEEE, vol. 86, no. 11, pp. 2278–2324 (1988)

43. Hubel, D., Wiesel, T.: Receptive fields and functional architecture of monkey striate cortex. J. Physiol. (London) **195**, 215–243 (1968)
44. CS231n Convolutional Neural Networks for Visual Recognition. http://cs231n.github.io/convolutional-networks/. Accessed 14 Sept 2017
45. Fukushima, K.: Neocognitron: a self-organizing neural network model for a mechanism of pattern recognition unaffected by shift in position. Biol. Cybern. **36**, 193–202 (1980)
46. Serre, T., Wolf, L., Bileschi, S., Riesenhuber, M.: Robust object recognition with cortex-like mechanisms. IEEE Trans. Pattern Anal. Mach. Intell. **29**(3), 411–426 (2007)
47. Ranzato, M., Poultney, C., Chopra, S., LeCun, Y.: Efficient learning of sparse representations with an energy-based model. In: Platt, J. et al. (eds.), Advances in neural information processing systems (NIPS 2006). MIT Press (2006)
48. Cireşan, D.C., Giusti, A., Gambardella, L.M., Schmidhuber, J.: Mitosis detection in breast cancer histology images with deep neural networks. In: Mori, K., Sakuma, I., Sato, Y., Barillot, C., Navab, N. (eds.) MICCAI 2013. LNCS, vol. 8150, pp. 411–418. Springer, Heidelberg (2013). https://doi.org/10.1007/978-3-642-40763-5_51
49. Sarraf, S., Tofighi, G.: Classification of Alzheimer's Disease using fMRI Data and Deep Learning Convolutional Neural Networks. arXiv preprint arXiv:1603.08631 (2016)
50. Liu, S., Liu, S., Cai, W., Che, H., Pujol, S., Kikinis, R., Adni, M.J.: Multi-modal neuroimaging feature learning for multi-class diagnosis of Alzheimer's disease. IEEE Trans. Biomed. Eng. **62**(4), 1132–1140 (2015). https://doi.org/10.1109/TBME.2014.2372011
51. van der Burgh, H.K., Schmidt, R., Westeneng, H.J., de Reus, M.A., van den Berg, L.H., van den Heuvel, M.P.: Deep learning predictions of survival based on MRI in amyotrophic lateral sclerosis. NeuroImage: Clinic. **13**, 361–369 (2017). ISSN 2213-1582. http://dx.doi.org/10.1016/j.nicl.2016.10.008
52. Hüsken, M., Stagge, P.: Recurrent neural networks for time series classification. Neurocomputing **50**, 223–235 (2003)
53. Pascanu, R., Mikolov, T., Bengio, Y.: Understanding the exploding gradient problem. Computing Research Repository (CoRR) abs/1211.5063 (2012)
54. Hochreiter, S., Schmidhuber, J.: Long short-term memory. Neural Comput. **9**(8), 1735–1780 (1997)
55. Graves, A., Mohamed, A.-R., Hinton, G.E.: Speech recognition with deep recurrent neural networks. In: IEEE International Conference on Acoustics, Speech and Signal Processing, pp. 6645–6649. IEEE (2013)
56. Breuel, T.M., Ul-Hasan, A., Al-Azawi, M.A., Shafait, F.: High-performance OCR for printed English and Fraktur using LSTM networks. In: 12th International Conference on Document Analysis and Recognition, pp. 683–687. IEEE (2013)
57. Gonzalez-Dominguez, J., Lopez-Moreno, I., Sak, H., Gonzalez-Rodriguez, J., Moreno, P.J.: Automatic language identification using long short-term memory recurrent neural networks. In: Proceedings of Interspeech (2014)
58. Geiger, J.T., Zhang, Z., Weninger, F., Schuller, B., Rigoll, G.: Robust speech recognition using long short-term memory recurrent neural networks for hybrid acoustic modelling. In: Proceedings of Interspeech (2014)
59. Fan, Y., Qian, Y., Xie, F., Soong, F.K.: TTS synthesis with bidirectional LSTM based recurrent neural networks. In: Proceedings of Interspeech (2014)
60. Barak, O.: Recurrent neural networks as versatile tool of neuroscience research. Curr. Opin. Neurobiol. **46**, 1–6 (2017)
61. Rajan, K., Harvey, C.D., Tank, D.W.: Recurrent network models of sequence generation and memory. Neuron **90**(1), 128–142 (2016). https://doi.org/10.1016/j.neuron.2016.02.009

62. Güçlü, U., van Gerven, M.A.J.: Modeling the dynamics of human brain activity with recurrent neural networks. Front. Comput. Neurosci. **11**, 7 (2017). https://doi.org/10.3389/fncom.2017.00007

63. Sussillo, D., Churchland, M.M., Kaufman, M.T., Shenoy, K.V.: A neural network that finds a naturalistic solution for the production of muscle activity. Nat. Neurosci. **18**, 1025–1033 (2015)

64. Vieira, S., Pinaya, W.H.L., Mechelli, A.: Using deep learning to investigate the neuroimaging correlates of psychiatric and neurological disorders: methods and applications. Neurosci. Biobehav. Rev. **74**, 58–75 (2017). https://doi.org/10.1016/j.neubiorev.2017.01.002

65. Donahue, J., Anne Hendricks, L., Guadarrama, S., Rohrbach, M., Venugopalan, S., Saenko, K., Darrell, T.: Long-term recurrent convolutional networks for visual recognition and description. In: Proceedings of the IEEE Conference on Computer Vision and Pattern Recognition, pp. 2625–2634 (2015)

# Computer Vision

# Investigating Crime Rate Prediction Using Street-Level Images and Siamese Convolutional Neural Networks

Virginia O. Andersson[✉], Marco A.F. Birck, and Ricardo M. Araujo

Federal University of Pelotas, Pelotas, RS, Brazil
vandersson@inf.ufpel.edu.br, virginia.andersson@gmail.com

**Abstract.** The analysis of the environment for crime prediction is based on the premise that criminal behavior is influenced by the nature of the environment in which occurs. Street-level images are the closest digital depiction available of the urban environment, in which most street crimes take place. This work proposes a crime rate prediction model that uses street-level images to classify street crimes into low or high crime rate levels. For that, we use a 4-Cardinal Siamese Convolution Neural Network (4-CSCNN) and train and test our analytic model in two regions of Rio de Janeiro, Brazil, that showed high street crime concentrations between the years of 2007 and 2016. With this preliminary experiment, we investigate the use of convolutional neural networks (CNN) for the task of crime rating through visual scene analysis and found possibilities towards automatic crime rate predictions using CNN models.

**Keywords:** Crime prediction · Computer vision · Street-level images · Siamese Convolutional Neural Networks

## 1 Introduction

Several works have been developed to understand the factors that trigger a criminal event carried out by an individual, as well as the risks involved and the measures to avoid it. Over the years, theories have been developed to map criminals behavior and crime itself. "Traditional criminological theories" are concerned about how developmental experiences, biological and social factors create a criminal offender [1]. On the other hand, "Environmental theories" consider crime as a confluence of *offenders*, *targets* and specific *laws and settings* at *particular times and places* [1,2]. In here, offenders are not the central object of interest, but one element of a *crime event*.

According to [1], the environmental perspective is based on the premise that criminal behavior is influenced by the nature of the environment in which occurs, i.e., the environment plays a fundamental role in initiating crime and shaping its course. The distribution of crime in time and space is non-random and it depends on environmental factors and situations. Control and crime prevention are then a result of understanding the role of the environment on crimes patterns.

© Springer International Publishing AG 2017
D.A.C. Barone et al. (Eds.): LAWCN 2017, CCIS 720, pp. 81–93, 2017.
https://doi.org/10.1007/978-3-319-71011-2_7

Environmental theories can focus on (i) how offenders react in the environment, such as the *Routine Activity* [3], the *Rational Choice* [4], and the *Crime Pattern Theory* [5], and (ii) how to map the physical environment in which the criminals operate, such as the well known *Broken Windows Theory* [6] and the *Routine Active of Places* [7].

Street-level images are the closest depiction of the human environment available in digital form, and their use in daily life is gradually increasing, mostly to aid navigation. The Google Street View [8] service popularized the access to such street images and computer vision models are being used together with street-level images to relate a city's physical appearance with crime statistics [9–11].

In this work, we present an initial study on crime rate prediction models that use street-level images as input. We propose a 4-Cardinal Siamese Convolution Neural Network (4-CSCNN) together with a Multi-layer Perceptron to classify visual scenes depicted in street-level images into low or high crime rates. We built a dataset with street-level images from two regions of Rio de Janeiro City, Brazil, that showed high street crimes rates between the years of 2000 and 2016. These regions were divided into equal sized cells containing the total crime events that happened inside the cell region. Images belonging to the cell received the label according to the total crimes in that region. We use this street-level images dataset to train and test our proposed model, obtaining 86% of overall accuracy in the classification of the two crime rate categories. The achieved results are discussed in Sect. 4.

This paper is organized as follows. Section 2 presents background and related work to Crime Prediction Models, Visual Scene and Environment Analysis and Siamese Convolutional Neural Networks. Section 3 describes how data was collected together with the methodology followed by our approach. Section 4 discusses the most important findings of our work, and Sect. 5 presents some conclusions and ideas for future work.

## 2   Related Work

Our paper is related to the following areas of academic literature: (i) Crime Prediction Models, that are the main purpose of this work, (ii) Visual Scene and Environment Analysis, which inspired the use of visual attributes in our proposed model and (iii) Siamese Convolutional Neural Networks, the deep learning technique which is used in our model.

### 2.1   Crime Prediction Models

Based on *Routine Activity of Places Theory* [7], crime *hot spots* are regions where high concentration of crime events is observed. The *hot spots* were initially used as a criminal data visualization technique, and further became a prediction model, with the advance of statistical and geographic information (GIS) tools, and the support of the observed characteristics of crime events *repeat* and *near-repeat* [7]. *Hot spots* were used by [12], in the *Early Warning System*, that used data collected by the community and law enforcement to produce the *hot spots* and point out possible near regions affected by violent crimes.

In [13], authors proposed the use of a technique to distribute crime in a geographic surface and calculate risk assessment of crime events - the prospective risk surface - to obtain *hot spots*. This technique consists of using a two-dimensional grid with $n$ equally sized cells overlying the geographic region of interest. In the model, a weight is associated with geographically located crimes. Recent crimes that happened near the center of a cell receives a higher weight. The weights of all crimes near the center are added together to produce the risk index of the belonging cell. To evaluate the proposed model, the authors used "theft" data from Merseyside County, England, from the year of 1997. The model presented 62 to 64% of accuracy in predicting crimes for a 2-day time window.

Later, [14] proposed the use of *Kernel Density Estimation* (KDE) to map *hot spots* overlaying a geographic area. It smooths the criminal data over a region according to a kernel density estimator function, mapping probabilities for a crime event happening under a specific area. In [15], the authors evaluate the KDE approach reaching 76 to 84% of accuracy in predicting crimes in a 3-month time window. More recently, [16] proposed a modification in KDE estimator to hold independent variables. The author used *tweets* from a specific location in Chicago city, Illinois, correlated with crime events in the location.

*Risk Terrain Modeling* (RTM) was proposed by [17] to assess crime risk over a region. RTM consist of acquiring crime-related factors and standardize each factor to a common geographic region, usually assigning a weight to the presence or absence of this factor at every place covered by the region of interest. As the risk value over a region gets higher, the probability of a crime event occurring in that region also gets higher. According to [17], the RTM technique produces maps that indicate regions with greater risks of becoming *hot spots* in the future. In [18], the authors propose an aggregate neighborhood risk of crime (ANROC) measure, applying RTM model to forecast neighborhood-level of violent crimes in Little Rock, Arkansas. They identify 14 risk factors, such as banks, bus stops, check-cashing, convenience stores, fast-food restaurants, grocery stores, hotel/motels, liquor stores, pawn shops, tattoo/piercing shops that were expected to influence crime. The ANROC measure was obtained by averaging the risk of crime per cell by neighborhood.

## 2.2   Visual Scene Environment Analysis

Computer vision and machine learning are used extensively to discover environment attributes in street-level images. According to [19], visual scene and urban imagery analysis can be focused on different objectives, such as predicting perceptual responses from images, understanding cities by their visual urban scene, understanding the connection between urban appearance and socioeconomic factors and rank or compare different urban environments. The fields in which this work is interested in are understanding cities and the connection between urban appearance and socioeconomic factors.

In [20], the authors proposed a methodology to automatically find visual elements (e.g. windows, balconies, street signs) from street-level images that are distinctive for a specific geographic area, i.e., they occur much more frequently in that area than other areas. The authors used Google Street View, a street-level image database, and clustered images using Histograms of Oriented Gradient (HOG) and color components to compose descriptors for specific patches (e.g. windows, signs, doors). They clustered the patches using Nearest Neighbor algorithm, dividing positive and negative data clusters into $l$ equally-sized subsets. Next, they iteratively trained an SVM detector for each visual element using the detectors trained in the previous iteration in a new unseen subset, selecting top $k$ detection for retraining. As results, for Paris, the authors achieved 83% of accuracy (where chance yields 50%) and for Prague City, 92% of accuracy.

Later, [9] followed the proposed model of [20] and applied Clustering and Support Vector Regression (SVR) to USA cities, discovering predictive relationships between visual elements from the environment and non-visual variables like crime and theft rates, housing prices, population density, graffiti density and perception of danger. They compared the use of HOG+color descriptors with the fifth convolutional layer of Caffe's ImageNet CNN model and concluded that HOG+color descriptors were more visually consistent but captured less city semantics.

In [10], the authors explored the ability to use visual scenes to predict distances of surroundings establishments such as hospitals and fast-food restaurants. They also explored the possibility of predicting crime rates in an area using visual scenes. The authors experimented different descriptors, e.g. GIST, Texture using Local Binary Pattern (LBP), Color using Locally-Constrained Linear Coding (LLC), HOG+Color and FC7 layers from Caffe's ImageNet CNN model. They used an SVR machine on the image features obtained by each descriptor. The results achieved in finding hospitals and fast-food restaurants showed similar results between Color, HOG+color and Deep Learning descriptors, varying from 0.58 to 0.61% of accuracy. To predict crime rates and danger perception, the authors used HOG+color descriptors and the SVR machine and compared to a human test prediction. The results achieved for crime rate prediction was an accuracy of 72.0%, compared to 59.6% in human tests.

### 2.3  Siamese Convolutional Neural Networks

In [21], Siamese Neural Networks are defined as "twin" Neural Networks (NNs) which share their parameters. The main reason to use twin NNs is that the inputs will be mapped to a very near space since they are being processed by the same function e.g. Convolutional Neural Networks (CNNs). The Siamese NNs are widely used to discriminate and match tasks, such as in [21–24].

Recently, focusing in city places safeness classification through street-level images, [19] proposed the Place Pulse 2.0, a novel dataset containing a pairwise comparison of "safe" and "unsafe" images from 56 cities across the world. The authors also proposed two different Siamese CNN architectures: Streetscore-CNN (SS-CNN), to predict a winner between two images comparison as safe and unsafe, and a Ranking Streetscore-CNN (RSS-CNN), where a ranking function is attached to the SS-CNN to rank the street-level images into high to low safeness scores.

A 4-Cardinal Directions pseudo-Siamese Convolutional Neural Network was proposed by [25], where the author used CNNs to extract a low-level representation of images from geographical points in Colorado, USA state, aiming at determining the location of the images. The CNNs architecture used was based on ResNet [26] and AlexNet [27], reserving two networks per cardinal direction, using the method of ensembles in conjunction. The network learns "from scratch" to classify regions from Colorado state. The representations mapped by the author's CNNs for each cardinal direction are concatenated and passed through two fully connected Multi-layer Perceptron layers (MLP) that classify the representation according to the given labels.

# 3    Methodology

We present in this work a preliminary model to predict crime rates in a city region, using only visual attributes extracted from street-level images from specific locations. The main objective of the model is, given a georeferenced location (point), predict the historical crime density in that location based on the images that surround it in the four cardinal orientations. We intend to understand how the visual scene influences the crime rate in each place.

## 3.1    Crime Data

The crime data used in this work was collected from the collaborative site *Wikicrimes* [28][1], where users can report criminal events from past and present. The *Wikicrimes* dataset contains records of crime events with latitude and longitude georeference, from different cities around the world, with most cities from Brazil. The dataset has criminal records between 2000 and 2016, and a few historical records from the 90's.

To visualize and identify crime hot spots in Rio de Janeiro city limits, depicted in Fig. 1(a), we create a grid-like data structure of shape $80 \times 80$, dividing the limits of the city into 6.400 equally spaced cells, and distribute the crime records into the corresponding cells, using latitude and longitude as reference, as presented in Fig. 1(b). Darker shades of blue are used to denote that at least one crime occurred in that area, and more crimes per cell are indicated by lighter and hotter shades of green, orange and red.

---

[1] http://www.wikicrimes.org.

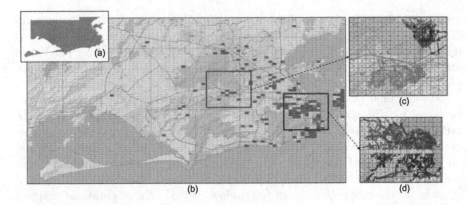

**Fig. 1.** In the detail (a): Rio de Janeiro city limits. In (b), the $80 \times 80$ grid with crime hot spots in Rio, between the year of 2007 and 2016, obtained from the collaborative site *Wikicrimes* [28]. In (c) the area with fewer crime rates and the location points contained in test (green) and train (magenta) datasets, and (d) the area with higher crime rates and the location points in the train (red) and test (blue) sets. The cells between the train and test datasets aren't used. (Color figure online)

### 3.2 Obtaining and Labeling Street-Level Imagery

For our study, we choose the areas of Rio de Janeiro with less crime rate concentration, in Fig. 1(c), and higher crime rate concentration, indicated in Fig. 1(d). From the selected regions, in Fig. 1(b), we obtained *shapefiles* from streets' distribution using *Mapzen Metro Extracts* service [29], superimposing the *shapefile* vectors on the map, as depicted in Fig. 1(c) and (d). The *shapefile* of Rio de Janeiro streets contains 2D points, or vertices, that shape the lines of the streets. We transformed the vertices into latitude and longitude coordinates and interpolated points between them.

With Google Street View API [8] we obtained 4 images, from the vertices and interpolated points that shape the streets, corresponding to the 4 cardinal directions. Each point, with their 4 images, was distributed by grid cell. Next, each cell from the selected area received a label according to the crime distribution in the region of interest. We categorized the distribution into "blue" and "red" to indicate low and high street crime rates.

Using the methodology described, we built a street-level imagery dataset with a total of 20,839 points, with 83,356 images for each point. Each point belongs to a specific cell and a specific label according to the crime rate in the cell region. To build the train and test dataset, we split the area into train and test regions, with one region containing cells and locations points for training and other region containing cells and location points for testing (Fig. 1(c) and (d)). We left some cells as a boundary, without using their locations points, to avoid training and testing with the same locations. Table 1 shows the total number of points for each label - blue and red - and the total number of images in the train set. The test set was adjusted to have the same number of points based on the smallest number of examples that one label had.

**Table 1.** Image dataset composition arranged by labels, in train and test set.

| Label | Train set | | Test set | |
|---|---|---|---|---|
| | Points | Images | Points | Images |
| Blue | 3,673 | 14,692 | 2,474 | 9,896 |
| Red | 12,218 | 48,872 | 2,474 | 9,896 |
| Total | 15,891 | 63,564 | 4,948 | 19,792 |

## 3.3 Proposed Siamese CNN

Our proposed model uses the concept of Siamese Neural Networks in which all the CNNs have their weights "frozen" and reused. This proposed preliminary architecture was inspired by the conjunction of ideas from [24, 25]. In [25], the author uses a variant of a Siamese network, named as a Pseudo-Siamese [24], where the CNNs used are free to learn specific features to each image subset.

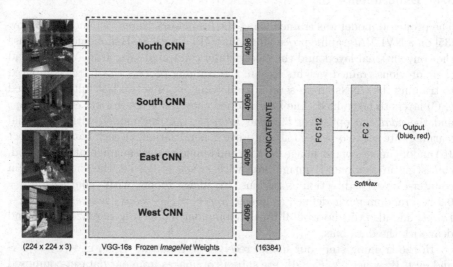

**Fig. 2.** Proposed 4-CSCNN architecture. The model receives as input 4 images corresponding to the 4 cardinal points of a given location, and each pre-trained VGG-16 network is responsible for extracting the best representative features of the images for the proposed problem during training step. Then, these features are concatenated into a single vector that is fed to a neural network, or fully connected layer, and classified into one of the established labels for criminality rate.

Figure 2 shows the proposed Siamese CNN architecture, showing the 4-cardinal CNNs with shared and frozen weights trained with *VGG-16* [30] using the *ImageNet* dataset [31], and the fully connected Multi-layer Perceptron (MLP) layers (FCs). The CNNs follows the VGG-16 architecture making our model more robust compared to the simpler CNN architecture used in [25]. Each CNN is frozen, following a common approach of *transfer learning*, allowing the use of pre-trained weights from the *ImageNet* domain. This way, the model is less likely to overfit and has the ability to leverage knowledge from the *ImageNet* domain.

In the proposed architecture, depicted in Fig. 2, each cardinal image from one location point is resized to $244 \times 244 \times 3$ pixels, and forwarded to each cardinal CNN. The resulting CNNs outputs are features vectors, with dimension of 4,096, from the first dense layer of the last Fully Connected (FC) block, right after the last Convolution layer, from the VGG-16 architecture. This feature vectors are then concatenated in a 16,384 feature vector and passed trough the classifier, an MLP that has been proved to be very efficient in conjunction with CNNs. The MLP has two fully connected layers, the *FC1 512* and *FC2 2*, where the backpropagation algorithm [32,33] is applied, and the output is passed to a *softmax* classifier to extract the distributed probability over the labels.

### 3.4   Experiments

The proposed model was trained using the framework Keras [34] and Tensorflow [35] on a NVIDIA graphic processing unit (GPU) with 8 GB of memory. Because the convolutional layers and the closest fully connected layer from VGG-16 had their previous trained weights frozen, the training process was faster compared to training the CNNs "from scratch", allowing only the two last fully connected (FC) layers to be trained. The loss function chosen was categorical *cross-entropy* and the *adaDelta* optimizer [36] was used to train the model, with the initial learning rate set to $\alpha = 2e-3$. To improve generalization, at each batch we select 16 random images of our image dataset and applied a data augmentation method, consisting in random (i) image cuts with $range = 0.2$ i.e. the shear angle in counter-clockwise direction as radians, (ii) image zooms, with $zoom - range = 0.2$ in a random mode defined as $[lower, upper] = [1 - zoom - range, 1 + zoom - range]$ and also (iii) horizontal flips, helping our model to generalize better and deal with the data bias.

In the training step, our model receives as inputs $n^i \in N$, $e^i \in E$, $s^i \in S$ and $w^i \in W$, where $N, E, S, W$ are subsets of images from our dataset composed of Google Street View images, taken from the cardinal directions North, East, South and West respectively. Additionally, $i$ is a geographical point of a street that belongs to a cell with a labeled crime rate. Each one of these four images was passed trough the model in batches of 16 location points for the mini-batch training. Each image linked to one cardinal direction was passed trough the respective CNN to obtain a representation in a low dimensional space, a descriptor with deep features of the image. All image descriptors are then concatenated to represent a new low-dimensional descriptor of the whole geographical point. Furthermore, the resulting descriptor is passed through the fully connected MLP layer whose objective is to map the geographic point to one of the two crime rate labels.

With the purpose of analyzing the difficulty of crime rate classification based on street-level images, and to further compare the results obtained from the proposed 4-CSCNN, we built a simplified model, replacing the CNNs by Histogram of Oriented Gradients (HOG) descriptors [37] and the fully-connected layers by a Linear Support Vector Machine Classifier (LinearSVC). We resize each image to $128 \times 128$ pixels and applied the HOG methodology using patches of $8 \times 8$ pixels, with 8 directions and 1 cell per block [38]. Each cardinal HOG resulted in a feature dimension of 2048. Similar to the proposed 4-CSCNN model, the features were concatenated, with the resulting dimension of 8192. Next, the LinearSVC algorithm was trained to classify each location point into low or high crime rate labels.

The baseline LinearSVC was trained with parameters $C = 1$, squared-hinge loss function and 1000 epochs, following standard parameters found in [39]. In order to observe the behavior of the baseline classifier in the task of classifying location points, we gradually increased the number of training examples, and validated the model in each configuration with 25% of the amount of training examples. We trained and tested the baseline model 10 times for the different training and test subsets, shuffling the datasets each time, and averaged the overall accuracy to obtain a score metric for further comparison.

# 4    Results and Discussion

Figure 3 shows the results obtained with the experiments performed using the Baseline HOG+LinearSVC classifier. In Fig. 3(a) the overall accuracy obtained for each subset of training and testing location points fairly increases from 100 training examples and 25 test examples, with an overall accuracy of 58 to 72% with 1000 training and 250 test examples. The accuracy starts to decrease as more examples are trained and tested, finally reaching the total number of examples of the dataset, obtaining 69% of overall accuracy.

The confusion matrix for the classification of all examples in test set using the Baseline HOG+LinearSVC is depicted in Fig. 3(b). More location points are classified as having high crime rates ("red" label) than low crime rates ("blue" label). In the baseline model, the probability that a location point will be classified as having a red label when it's a high crime rate area is 84.7% (sensitivity), and the probability that the model will classify a location point with a blue label when it's a low crime rate area is 54.6% (specificity).

Figure 4 shows the results obtained when training and testing our proposed model 4-CSCNN with VGG-16 pre-trained weights using the *ImageNet* dataset. The model reaches reasonable results in the first 20th epochs, with 80% of overall accuracy score, and 0.4 of loss value, i.e., the sum of the errors obtained during the evaluation of each example, expected to be low during test step. As the location examples are presented in each epoch, the overall accuracy varies in small amounts, as depicted in Fig. 4(a), until the 180th epoch, when the model reaches 86% of overall accuracy during test step, with a loss value of 0.3.

The classification of the examples in the test set, at the 180th epoch, resulted in the confusion matrix depicted in Fig. 4(b). Similar to the baseline model, the

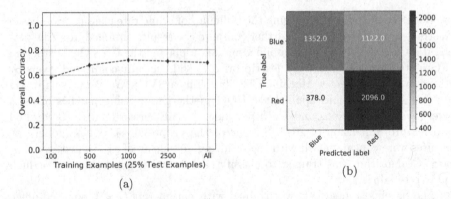

**Fig. 3.** In (a) overall accuracies obtained with the Baseline HOG+LinearSVC classifiers, varying the number of training examples and test examples, taken as 25% of the training examples. In (b) the resulting confusion matrix from the classification of 4948 location points using all training examples available. (Color figure online)

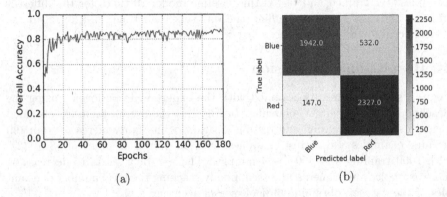

**Fig. 4.** In (a) overall accuracies obtained with the 4-CSCNN proposed architecture, during training and testing in 180 epochs. In (b) the resulting confusion matrix from the classification of 4948 location points using all training examples available at the 180 epoch. (Color figure online)

proposed 4-CSCNN model predicted more locations as being a high crime rate area, or "red", then being a low crime rates area, or "blue". Considering the "red" as positive and "blue" as negative, relative to the presence or absence of crime, the probability of the model in predicting a high crime locality as a high crime rate area, or "red", is 94% (sensitivity), and the probability of the proposed model in predicting a locality with low crime rates as being "blue" is 78.5%.

Table 2 shows the results achieved with the Baseline HOG+LinearSVC model, compared with the results from the proposed 4-CSCNN network. The results obtained with the simplified model demonstrate the non-trivial nature of the problem addressed, necessitating the investigation of a model that better

**Table 2.** Comparison between the proposed 4-CSCNN and the Baseline model with HOGs and LinearSVC classifier.

| Model | Overall accuracy | True positive rate | True negative rate |
|---|---|---|---|
| HOG+LinearSVC | 69% | 84.7% | 54.6% |
| **4-CSCNN** | **86%** | **94%** | **78.5%** |

serves the classification of the criminal rate through street-level images. Our proposed 4-CSCNN performed better than the Baseline HOG+LinearSVC model, considering the overall accuracy and the true positive rate achieved.

# 5   Conclusion

This paper presents a preliminary study on predicting crime rates from street-level images, which represent the urban environment where street crime occurs. For this, we proposed a new 4-Cardinal Siamese Convolutional Neural Network (4-CSCNN) architecture to predict urban crime rates, given a georeferenced location point. The model uses 4 images surrounding the given point, facing north, south, east and west. Each image is the input of one CNN, with pre-trained frozen weights from *VGG-16* architecture [30], trained with the *ImageNet* dataset [31]. At the output of each CNN, a *Fully Connected* (FC) layer was attached, and the resulting descriptors were merged into one only descriptor, that is finally classified by a Multi-layer Perceptron (MLP) into one of the two crime rate labels: "blue", indicating low crime rate and "red", indicating high crime rates.

The CNNs are responsible for learning features of the environment images which may affect crime rates. The use of 4 images surrounding a location gives more information about the environment than using a single image as input. Considering the overall accuracy score, the proposed 4-CSCNN achieved better results when compared with a simple baseline model, using HOG descriptors and a Linear Support Vector Classifier. The obtained results, 86% of overall accuracy, indicates that the architecture can infer a possible relation between environment features and crime rates, using only street-level images.

For future works, we intend to implement techniques of deep visualization in the neurons, to show what image inputs cause higher activation in units. By doing this, we search for the features that the CNNs learned to be related to low or high crime rates. This can be useful for social and law enforcement analysis of the urban environment. Also, our model has potential to be applied to different problems i.e. 4-Cardinal images of one georeferenced point can be related to different statistics and environment characteristics, e.g. city region classification as downtown and suburbs, that requires surrounding visualization.

**Acknowledgemnts.** We gratefully acknowledge the support of NVIDIA Corporation with the donation of the Titan X GPU used for this research. This work is supported by the Research Initiation Scholarship Program - Doctorate in Progress (PBIP-DA) from Federal University of Pelotas (UFPel).

# References

1. Wortley, R., Mazerolle, L.: Environmental criminology and crime analysis: situating the theory, analytic approach and application (2008)
2. Brantingham, P.L., Brantingham, P.J.: Criminality of place. Eur. J. Crim. Policy Res. **3**(3), 5–26 (1995)
3. Cohen, L.E., Felson, M.: Social change and crime rate trends: a routine activity approach. Am. Sociol. Rev. **44**(4), 588–608 (1979)
4. Brantingham, P.J., Brantingham, P.L.: Environment, routine and situation: toward a pattern theory of crime. Adv. Criminolog. Theory **5**, 259–294 (1993)
5. Eck, J.E., Weisburd, D.L.: Crime places in crime theory. Crime Prevent. Stud. **4**, 1–33 (1995)
6. Wilson, J.Q., Kelling, G.L.: Broken windows. Atl. Month. **249**(3), 29 (1982)
7. Sherman, L.W., Gartin, P.R., Burger, M.E.: Hot spots of predatory crime: routine activitices and the criminology of place. Criminology **27**(June), 27–55 (1989)
8. Google: Google Street View API (2017). https://developers.google.com/maps/documentation/streetview. Accessed 09 May 2017
9. Arietta, S.M., Efros, A.A.: City forensics: using visual elements to predict non-visual city attributes. Trans. Visual. Comput. Graph. **20**(12), 2624–2633 (2014)
10. Khosla, A., An, B., Lim, J.: Looking beyond the visible scene. In: 2014 IEEE Conference on Computer Vision and Pattern Recognition (2014)
11. Gebru, T., Krause, J., Wang, Y., Chen, D., Deng, J., Aiden, E.L., Fei-Fei, L.: Using deep learning and Google street view to estimate the demographic makeup of the US, pp. 1–41 (2017)
12. Block, C.: The GeoArchive: an information foundation for community policing. In: Crime Mapping and Crime Prevention, pp. 27–81 (1998)
13. Bowers, K.J., Johnson, S.D., Pease, K.: Prospective hot-spotting: the future of crime mapping? Br. J. Criminol. **44**(5), 641–658 (2004)
14. Chainey, S., Tompson, L., Uhlig, S.: The utility of hotspot mapping for predicting spatial patterns of crime. Secur. J. **21**, 4–28 (2008)
15. Johansson, E., Gahlin, C., Borg, A.: Crime hotspots: an evaluation of the KDE spatial mapping technique. In: EISIC European Intelligence and Security Informatics Conference, Manchester, UK, pp. 69–74. IEEE (2015)
16. Gerber, M.S.: Predicting crime using Twitter and kernel density estimation. Decis. Support Syst. **61**(1), 115–125 (2014)
17. Caplan, J.M., Kennedy, L.W., Miller, J.: Risk terrain modeling: brokering criminological theory and GIS methods for crime forecasting. Justice Q. **28**(2), 360–381 (2011)
18. Drawve, G., Thomas, S.A., Walker, J.T.: Bringing the physical environment back into neighborhood research: the utility of RTM for developing an aggregate neighborhood risk of crime measure. J. Crim. Justice **44**, 21–29 (2016)
19. Dubey, A., Naik, N., Parikh, D., Raskar, R., Hidalgo, C.A.: Deep learning the city: quantifying urban perception at a global scale. pp. 196–212 (2016)
20. Doersch, C., Singh, S., Gupta, A., Sivic, J., Efros, A.: What makes Paris look like Paris? ACM Trans. Graph. **31**(4), 1–9 (2012)

21. Bromley, J., Bentz, J.W., Bottou, L., Guyon, I., LeCun, Y., Moore, C., Säckinger, E., Shah, R.: Signature verification using a "siamese" time delay neural network. Int. J. Pattern Recogn. Artif. Intell. **7**(04), 669–688 (1993)
22. Taigman, Y., Yang, M., Ranzato, M., Wolf, L.: Deepface: closing the gap to human-level performance in face verification. In: Proceedings of IEEE Conference on Computer Vision and Pattern Recognition, pp. 1701–1708 (2014)
23. Lin, T.-Y., Cui, Y., Belongie, S., Hays, J.: Learning deep representations for ground-to-aerial geolocalization. In: 2015 IEEE Conference on Computer Vision and Pattern Recognition (CVPR), pp. 5007–5015. IEEE (2015)
24. Zagoruyko, S., Komodakis, N.: Learning to compare image patches via convolutional neural networks. In: Proceedings of IEEE Conference on Computer Vision and Pattern Recognition, pp. 4353–4361 (2015)
25. Lieman-Sifry, J.: Convolutional neural networks to predict location from Colorado Google street view images: Galvanize capstone project (2016). https://github.com/jliemansifry/streetview/
26. He, K., Zhang, X., Ren, S., Sun, J.: Deep-residual learning for image recognition. arXiv preprint arXiv:1512.03385 (2015)
27. Krizhevsky, A., Sutskever, I., Hinton, G.E.: ImageNet classification with deep convolutional neural networks. In: Advances in Neural Information Processing Systems, pp. 1097–1105 (2012)
28. Furtado, V., Ayres, L., de Oliveira, M., Vasconcelos, E., Caminha, C., D'Orleans, J., Belchior, M.: Collective intelligence in law enforcement - the WikiCrimes system. Inf. Sci. (NY) **180**(1), 4–17 (2010)
29. Mapzen: Mapzen metro extracts (2017). https://mapzen.com/data/metro-extracts/. Accessed 09 May 2017
30. Simonyan, K., Zisserman, A.: Very deep convolutional networks for large-scale image recognition. ImageNet Challenge, vol. 110 (2014). https://doi.org/10.1016/j.infsof.2008.09.005
31. Deng, J., Dong, W., Socher, R., Li, L.-J., Li, K., Fei-Fei, L.: ImageNet: a large-scale hierarchical image database. In: 2009 IEEE Conference on Computer Vision and Pattern Recognition, pp. 248–255 (2009)
32. Rumelhart, D.E., Hinton, G.E., Williams, R.J.: Learning representations by back-propagating errors. Cogn. Modeling **5**(3), 1 (1988)
33. LeCun, Y., Bottou, L., Bengio, Y., Haffner, P.: Gradient-based learning applied to document recognition. Proc. IEEE **86**(11), 2278–2324 (1998)
34. Chollet, F.: Keras (2015). https://github.com/fchollet/keras
35. Abadi, M., Agarwal, A., Barham, P., Brevdo, E., Chen, Z., Citro, C., Corrado, G.S., Davis, A., Dean, J., Devin, M., Ghemawat, S.: TensorFlow: large-scale machine learning on heterogeneous systems (2015). Software available from tensorflow.org
36. Zeiler, M.D.: Adadelta: an adaptive learning rate method. arXiv preprint arXiv:1212.5701 (2012)
37. Dalal, N., Triggs, B.: Histograms of oriented gradients for human detection. In: Proceedings - 2005 IEEE Computer Society Conference on Computer Vision and Pattern Recognition, CVPR 2005, vol. I, pp. 886–893 (2005)
38. van der Walt, S., Schönberger, J.L., Nunez-Iglesias, J., Boulogne, F., Warner, J.D., Yager, N., Gouillart, E., Yu, T.: Scikit-image: image processing in Python. PeerJ **2**, e453 (2014)
39. Pedregosa, F., Varoquaux, G., Gramfort, A., Michel, V., Thirion, B., Grisel, O., Blondel, M., Prettenhofer, P., Weiss, R., Dubourg, V., Vanderplas, J., Passos, A., Cournapeau, D., Brucher, M., Perrot, M., Duchesnay, E.: Scikit-learn: machine learning in Python. J. Mach. Learn. Res. **12**, 2825–2830 (2011)

# Developing of a Video-Based Model for UAV Autonomous Navigation

Wilbert G. Aguilar[1,3,4(✉)], Vinicio S. Salcedo[2,3],
David S. Sandoval[2,3], and Bryan Cobeña[2,3]

[1] Departamento de Seguridad y Defensa DESP, Universidad de las Fuerzas
Armadas ESPE, Sangolquí, Ecuador
wgaguilar@espe.edu.ec
[2] Departamento de Eléctrica y Electrónica DEEE, Universidad de las Fuerzas
Armadas ESPE, Sangolquí, Ecuador
{vssalcedo,dssandoval,gbcobena}@espe.edu.ec
[3] Centro de Investigación Científica y Tecnológica del Ejército CICTE,
Universidad de las Fuerzas Armadas ESPE, Sangolquí, Ecuador
[4] Research Group on Knowledge Engineering GREC, Universitat Politècnica de
Catalunya UPC-BarcelonaTech, Barcelona, Spain

**Abstract.** The use of unmanned aerial vehicles (UAVs) has driven the research
and development of multiple applications. Autonomous and cognitive navigation in remote environments requires the use of independent on board sensors.
One advantage of these vehicles is that they have an on-board camera that
allows them to collect visual information about the environment. This work
shows a way to be aware of the UAV movement depending only on images.
Therefore, a vision-based mathematical model was defined that describes the
movement. System identification experiments and results are presented to verify
the mathematical model structure and to identify model parameters comparing
with state of art models. Finally a visual-based model compare with other
methods and improve performance.

**Keywords:** Keypoints · Affine transform · Matching · UAV · Subscriber ·
Mathematical model

## 1 Introduction

The unmanned aerial vehicles have shown a significant demand with a market
increasing [1] which grows over time as assumed [2], [3]. Today UAV's are common
on air both in military applications as in public service, surveillance, monitoring, and
others [1]; where micro UAV's take the lead for being able for flying in closed and
small environments with multiple sensors and cameras that allow the implementation of
a control system and strengthen the navigation of UAV's, control systems as systems
based on the images from the devices of vision on board, getting closer to the goal of
UAV's to improve the autonomy and robustness to different climatic conditions.

Vision-based navigation systems depend mainly on the correct image capture and
heuristic selection [4, 5], however, when taking pictures on mobile platforms the

© Springer International Publishing AG 2017
D.A.C. Barone et al. (Eds.): LAWCN 2017, CCIS 720, pp. 94–105, 2017.
https://doi.org/10.1007/978-3-319-71011-2_8

motion dynamics can generate undesirable effects on images [6], translation and rotation effects due to platform instability, important on UAV's. For this, large companies that commercialize UAV's have opted for some methods to reach the video-stability in the flight as the servo visual control and software algorithms for image-captured correction and consecutive frames stabilization [7–9].

2D images taken from the on-board capture device of the UAV can be related to 3D geometry by an Euclidean model of perspective projection in image formation process as a set in $R^3$ that is point-to-point related with the Euclidean space $E^3$ [4, 5], all based on pin hole camera model where a projection of 3D space is done in a point-to-point plane passing through a hole or projection center [10–13].

The goal on this work is to estimate the motion model of the Parrot Bebop drone only based on captured images from on-board camera. For which an image processing is integrated by keypoints detection, description and matching. Then Random Sample Consensus (RANSAC) [14] is used to remove outliers. Finally, affine transformations to determine the behavior of the system based on frame to frame geometric transformations and its subsequent modeling.

This paper is organized as follows: Sect. 2 presents a description of the work system structure, Sect. 3 presents the used for image processing and extraction of geometric parameters methodology, Sect. 4 deals with the subject of modeling with record data from image processing; experimental results and system conditions, in addition to their comparison in performance with traditional methods are presented in Sect. 5 the conclusions and future work are presented in the last section.

## 2 System Overview

The present work is developed on Bebop mini drone of Parrot brand, an UAV to the micro scale in the VI classification according [15] which is easy to assemble and with ten minutes of autonomy in the air, also has a Powerful built-in computer, flight and video stability algorithms, flight safety and homecoming that are essential to prevent emerging events.

Most important for the present project is that Bebop drone has a 14-megapixel front-facing fisheye camera (180°) and unique three-axis image stabilization algorithms. This enable to obtain high quality images over a wide range and without distortions generated on the flight.

The system is composed by the mini drone which has the vision device incorporated. It is controlled from a ground station from which orders are sent, where the information provided during the flight is processed (Fig. 1).

### 2.1 Image Extraction

The algorithm used in this work accesses to images taken with the camera on board the drone using ROS. The *autonomy_bebop* driver based on the official Parrot *ARDrone_SDK3* SDK. This allows the usage of various functions of the drone by translating them into topics, services, messages and parameters. This driver allows to control the main functions like lift flight, landing, navigate, and more through the corresponding

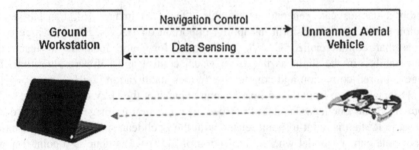

**Fig. 1.** System structure for a video-based model

topics (takeoff, land, cmd_vel, etc.). The images captured by the drone are published in the *image_raw* topic as a message of type *sensor_msgs/Image* and the calibration data in the *camera_info* topic.

One of most important points of Bebop Parrot ARDroneSDK3 is that video quality is limited to 640 × 368 image at 30 Hz. It allows to execute a virtual camera with which the capture position can be varied between about 80° horizontal and 50° vertical.

In ROS system, a topic is responsible for interconnecting the functions of the system. The obtaining images process published in a topic is called image subscribers, this process is necessary since ROS publishes the images as message type and if it is desired to perform a post processing it must be converted to OpenCV images and corresponds to the diagram presented in Fig. 2.

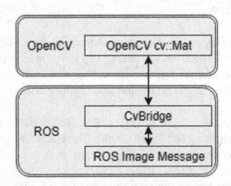

**Fig. 2.** Image conversion between ROS message type and OpenCV.

## 3    Methodology

### 3.1    Image Features

To perform the UAV described motion analysis, it is necessary to extract the characteristics of the images obtained during the flight, then compare them and determine the system behavior. This procedure includes image processing algorithms that perform the needed operations to get these characteristics.

These concepts are based on image analysis as an intensity matrix in each pixel. The intensity changes in the image (applied in edge detection [9]) are identified by a threshold and a hysteresis level, and determine the connection between the keypoints at detected edges.

The image characterization is defined by three phases: firstly, the keypoints must be detected, then the characteristic descriptor vectors must be determined, and finally, they must be compared and matching, allowing the appreciation of the similarities between frames.

**Keypoints detection.** One of main advantages when using algorithms for detection of keypoints is them low computational cost [16], there are multiple algorithms that allow the keypoints detection; Harris detectors [17], for example, choose to find the fastest and slowest direction changes for the targeting characteristics. This is solved by a covariance matrix of local directional derivatives resulting in high rotational invariance and robustness to intensity changes. Other high-efficiency detectors are SIFT (Scale Invariant Feature Transform [18–20]) that use Gaussian difference [17] to determine possible points, simulating a values filtering that are then compared in the neighborhood of $3 \times 3 \times 3$. Thus the differences are classified and the points are determined. It is worth noting that this one has a high invariance to changes of translation, rotation and scaling allowing a good keypoints detection in noisy environments as the drone case.

The calculation speed of SIFT method is overshadowed by Speeded Up Robust Features (SURF), which is ten times faster. This is based on an approximation of the Hessian matrix as an analysis of the maximum in this determinant (Process detailed in [16]). However, there is an algorithm that can give a better performance solution called ORB (Oriented FAST and Rotated BRIEF [21–23]). This is based on the FAST key-point detector and BRIEF descriptors. This allows the real time keypoints detection by taking a parameter of intensity and comparing between the central pixel and the circle of nine neighbors. ORB acquires its order by means of a Harris corners measurement [17] for the selection of the superiors in a pyramid of scale points of the image.

**Keypoints description.** The descriptors are the way to have in few data the image features, basing their philosophy in being a complete, compact, invariant and congruent set of data [9].

As mentioned above the ORB descriptor is BRIEF based but additionally it is added a rotational invariance giving it the orientation by the FAST9 use followed by a metric of Harris. This orientation is calculated by the use of intensity centroid with a sampling pattern of 256 intensity comparisons in pairs. Binary patterns are built through automatic learning [17].

**Keypoints matching.** Consequently, to get the descriptors of each keypoint found in the image, the subsequent process is matching, then correspondences between consecutive frame points are found. We compare the descriptors of each as a vector distance, showing an error corresponding to the sample that differs less between points in the frame.

To improve the performance of this process, we has been integrated an algorithm based on the minimization of gray level difference called RANSAC (Random Sample Consensus). Which remove the false correspondences that occur in the process and also the uncertainties in the geometric analysis.

In Fig. 3 we can see the result of applying the algorithm for the matching followed by RANSAC. Getting the best correspondences between the keypoints in the frames sequence. For the algorithm test a video of 80 frames of a circle movement was taken, next some of the resulting ones are shown

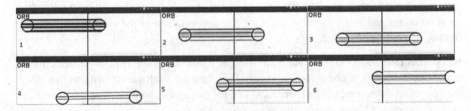

**Fig. 3.** Matching of a video sequence for algorithm demonstration.

## 3.2 Geometric Analysis

Once the correspondences between keypoints have been obtained, the geometric analysis of the images is carried out. This requires a process that relates 3D space to captured 2D, since the drone motion will be captured only by on board camera.

For navigation, the range of drone movements are on x-y plane (pitch and roll movements) are limited, as shown in Fig. 4.

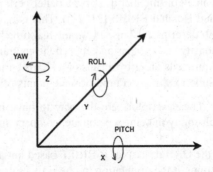

**Fig. 4.** Axes in the plane with the associated yaw, pitch and roll movements.

It is assumed that the x-y plane is ground parallel for the drone.

Another important factor is the need to capture images of the floor (ground plane) to make the geometric analysis more efficient. The 180° vision has been taken advantage directing the frame in sections of the lowest allowable take from autonomy_bebop driver, obtaining images of the ground plane at a fixed height.

In order to extract the motion parameters, necessary for the modeling of the system, a transformation that relates the space in R3 to a 2D approach is used. The affine transformation is used, since in small and flat surfaces the change of view of an image can be represented as an affine distortion with an associated scale.

The affine matrix has translation, rotation and scale parameters. The following is a multiplication and addition matrix where a point x1 is transformed into x2 by means of affine matrix:

$$\begin{bmatrix} x_2 \\ y_2 \end{bmatrix} = \begin{bmatrix} cos(\emptyset) & sin(\emptyset) \\ -sin(\emptyset) & cos(\emptyset) \end{bmatrix} \begin{bmatrix} s_x \\ s_y \end{bmatrix} \begin{bmatrix} x_1 \\ y_1 \end{bmatrix} + \begin{bmatrix} t_x \\ t_y \end{bmatrix} \tag{1}$$

where $\emptyset$ is the rotation angle $s_{x,y}$ are the scales and $t_{x,y}$ the translations on their respective axis y can be expressed more generally as:

$$X_2 = ASX_1 + T \tag{2}$$

The reason why this transformation is going to be used and not homography is because the affine transformation is a special case of the homography matrix. Where z = 1 [9], i.e. the plane is limited to Euclidean points in a fixed plane and is represented as a homography matrix as follows:

$$X_2 = \begin{bmatrix} A & T \\ 0 & 1 \end{bmatrix} X_1 \tag{3}$$

So that it adapts perfectly to presented problem, trying to capture the motion parameters. Mainly in translation of the consecutive images, to later carry out the estimation of the drone motion mathematical model.

As we can see in Fig. 5 the images geometric processing has been related to the translation parameters in the (x, y) axes for the drone. Shown the path traced for the sequence of movements of the previous figure.

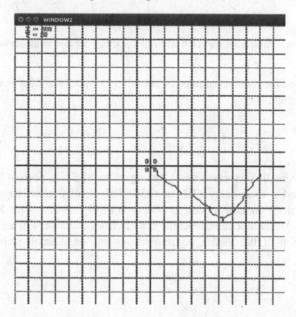

**Fig. 5.** Video sequence geometric translation of Fig. 4, getting by image processing

## 4 Image Mathematical Model

To get a drone motion mathematical model estimate, a video was taken aboard a fixed movement's sequence. These respond to control commands sent from ground station. The control actions that allow to perform the driver are: the movement of the virtual camera, movements in yaw, pitch, roll and altitude.

For this model has been determined a fixed height for the image capture, also pitch and roll movements exclusively to cause changes in the x-y plane.

To perform the tests, a pulse train was sent with the drone different speeds permissible in pitch and roll on the scale of [0–1] (Fig. 6). The variation has a resolution of 0.1 between each pulse. As result of the generated image processing, we get the data of a displacement delta which will be used for the system modeling (Fig. 7).

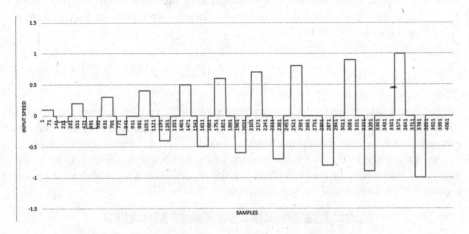

**Fig. 6.** Pulse train for pitch and roll velocities applied to the drone.

Once captured the motion parameters, the Matlab model identifier is used to obtain the transfer function:

$$G(s) = \frac{K_p}{1 + T_p \times s} \times e^{-T_D \times s} \tag{4}$$

Of which there is a process gain (Kp) equal to 82.72, a process time constant (Tp) of 0.83 and a time delay (Td) of 0.53. This results in an appreciably good approximation of the motion model, as observed in Figs. 8 and 9 for Roll and Yaw respectively. The black line represents the actual values of movement and the blue line the approximation of the model, additionally counts on a small stabilization time (Fig. 10).

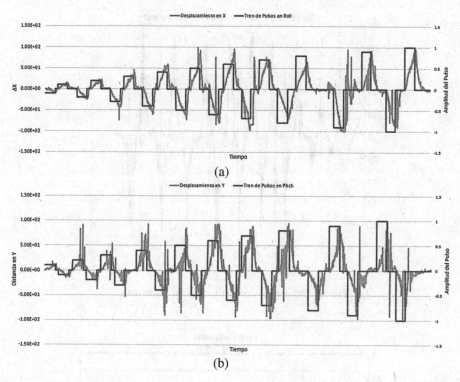

(a)

(b)

**Fig. 7.** (a) Drone behavior in X axis by Roll movements (b) Drone behavior in Y axis by Pitch movements.

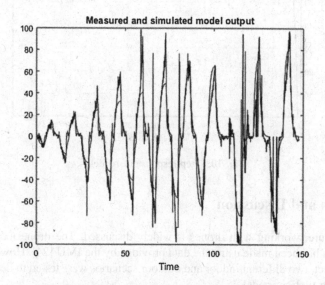

**Fig. 8.** Adjusting the motion model for Roll data.

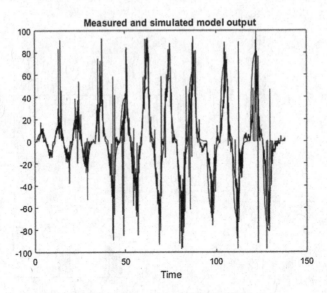

**Fig. 9.** Setting the motion model for Pitch data

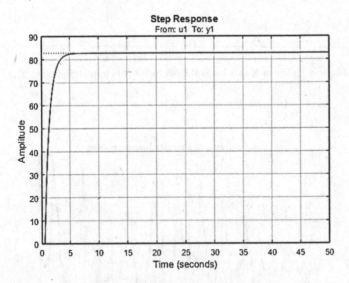

**Fig. 10.** Step response of model.

## 5  Results and Discussion

In the literature, working with drones is widely discussed. The drone motion model identification has been made using the data provided by the IMU [24]. However, in the present project, two different indoor and outdoor scenarios were tested to determine the differences in motion model.

Our approach requires several conditions for its correct functioning among which we must mention:

- The drone must be able to capture video directly to the floor so that the affine transformation has a high performance. As mentioned above is very useful in case of surfaces with flat and variable appearance.
- The surface on which the tests can be performed must have irregularities or contrast changes. The model is based on keypoints capture, these points cannot be captured on smooth surfaces without contrast changes.
- The video capture rate must be at least 30 fps in order to have enough data and decrease the change between frames that would cause major errors.

This method is mainly useful in case of UAV's that doesn't have an IMU good performance or is difficult to access it. In Bebop case it does not provide navdata as did its predecessor AR Drone Parrot sending navigation data at so high frequencies, otherwise Bebop only delivers data at an estimated frequency of 5 Hz which is not sufficient to assimilate a true drone motion model.

One of the problems that must be addressed in this approach is the geometric transformation correct response since it depends on the position and correct keypoints capture. The affine transformation is performed with three pairs of points so that it can be estimated Geometric difference between them. In process for keypoints detection can be obtained many points, in that case it should be chosen the best located. Since the estimation affine has less effectiveness on near points or when the lines between these points has similar slope. To address this problem, we used statistical methods for correct points selection and probability density. We determine the correct geometric values of those obtained from a set of affine transformations applied to the same image but with different points.

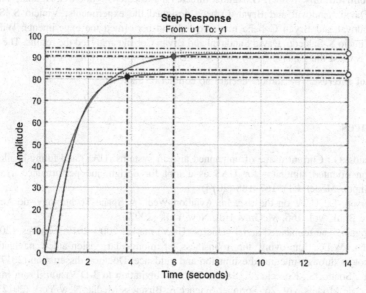

**Fig. 11.** Response time of the model using images (Blue), model response time using IMU (Red) (Color figure online)

In the work carried out by [24] they make an estimate of the motion model in Roll of a Parrot Bebop through the data provided by the IMU, nevertheless the response of its model ends up being slower than the one obtained in our proposal. Figure 11 shows a comparison of the response in both models and the stabilization time.

# 6  Conclusions and Future Works

In this paper we have shown the process of getting an approximation of the motion model for a drone. This system is very useful for UAVs with good video capture capability. The approach is oriented to the autonomous navigation of a drone addressing the problem of losing the GPS signal and experiencing IMU malfunction. It gives an option for the continuous modeling of UAVs with low computational cost and only depending on the camera on board.

The use of affine transforms to determine the geometric change between frames with flat distortion characteristics have well performs if takes the correct transformation points between frames. If the frames are consecutive the points are going to vary making of this a statistical process but with favorable results.

For future work, we pretend to design an algorithm to model a drone based on camera. That is not on board but in another plane using keypoints and affine transformations to geometrically relate the images and obtain the patterns of movement.

**Acknowledgments.** This work is part of the project "Perception and localization system for autonomous navigation of rotor micro aerial vehicle in GPS-denied environments (VisualNav Drone)" from the Centro de Investigación Científica y Tecnológica del Ejército (CICTE), directed by Wilbert G. Aguilar.

**Author Contributions.** Wilbert G. Aguilar directed the research; Wilbert G. Aguilar, Vinicio S. Salcedo, David Sandoval and Bryan Cobeña designed the experiments; Vinicio S. Salcedo, David Sandoval and Bryan Cobeña implemented and performed the experiments; Wilbert G. Aguilar, Vinicio S. Salcedo, David Sandoval and Bryan Cobeña analyzed the results. The authors wrote and revised the paper

**Conflicts of Interest.** The authors declare no conflict of interest.

# References

1. Limnaios, G.: Current usage of unmanned aircraft systems (UAS) and future challenges: a mission oriented simulator for UAS as a tool for design and performance evaluation. J. Comput. Model **4**(1), 167–188 (2014)
2. Dickerson, L.: UAV on the rise. In: Aviation Week & Space Technology. de Aerospace Source Book, vol. 166, McGraw Hill, New York (2007)
3. Visiongain: The unmanned aerial vehicles (UAV) market 2009–2019. Londres, (2009)
4. Whitten, W.D.: Improving the robustness of monocular vision-aided navigation for multirotors through integrated estimation and guidance. Doctoral dissertation (2017)
5. Ma, Y., Soatto, S., Kosecka, J., Sastry, S.S.: An Invitation to 3-D Vision: From Images to Geometric Models, vol. 26. Springer Science & Business Media, New York (2012)

6. Wilbert, G., Aguilar, C.A.: Compensación y Aprendizaje de Efectos Generados en la Imagen durante el Desplazamiento de un Robot. In: X Simposio CEA de Ingeniería de Control, pp. 165–170 (2012)
7. Aguilar, W.G., Angulo, C.: Real-time model-based video stabilization for microaerial vehicles. Neural Process. Lett. **43**(2), 459–477 (2016)
8. Guenard, N., Hamel, T., Mahony, R.: A practical visual servo control for an unmanned aerial vehicle. IEEE Trans. Robot. **24**(2), 331–340 (2008)
9. Aguilar, W., Angulo, C.: Robust video stabilization based on motion intention for low-cost micro aerial vehicles. In: A: International Multi-conference on Systems, Signals and Devices. B: Proceedings of 11th International Multi-conference on Systems, Signals and Devices (SSD 2014). Castelldefels: 2014, pp. 1–6 (2014)
10. Ganci, S.: Looking through a pinhole: physical and physiological phenomena. J. Photon. Opt. Technol. **3**(2), 13–16 (2017)
11. Wu, A., Johnson, E.N., Kaess, M., Dellaert, F., Chowdhary, G.: Autonomous flight in GPS-denied environments using monocular vision and inertial sensors. J. Aerospace Inf. Sys. **10**(4), 172–186 (2013)
12. Chowdhary, G., Johnson, E.N., Magree, D., Wu, A., Shein, A.: GPS-denied indoor and outdoor monocular vision aided navigation and control of unmanned aircraft. J. Field Robot. **30**(3), 415–438 (2013)
13. Nixon, M., Aguado, A.S.: Feature Extraction & Image Processing, 2nd edn. Academic Press, Cambridge (2008)
14. Derpanis, K.G.: Overview of the RANSAC algorithm. Image Rochester NY **4**(1), 2–3 (2010)
15. Kendoul, F.: Survey of advances in guidance, navigation, and control of unmanned rotorcraft systems. J. Field Robot. **29**(2), 315–378 (2012)
16. Wilbert, G., Aguilar, C.A.: Compensación de los efectos generados en la imagen por el control de navegación del robot Aibo ERS 7. In: Conference: VII Congreso de Ciencia y Tecnología ESPE, At Sangolquí (2012)
17. Sayem, A.S.S.: Vision-Aided Navigation for Autonomous Vehicles Using Tracked Feature Points (2016)
18. Zhang, X., Wang, X., Yuan, X., Wang, S.: An improved SIFT algorithm in the application of close-range stereo image matching. In: IOP Conference Series: Earth and Environmental Science, vol. 46, no. 1, p. 012009. IOP Publishing (2016)
19. Al-khafaji, S.L., Zhou, J., Zia, A., Liew, A.W.C.: Spectral-spatial scale invariant feature transform for hyperspectral images. IEEE Trans. Image Process. (2017)
20. Lowe, D.: Object recognition from local scale-invariant features. In: Proceedings of IEEE International Conference on Computer Vision, vol. 2, pp. 1150–1157 (1999)
21. Zhu, Y., Shen, X., Chen, H.: Copy-move forgery detection based on scaled ORB. Multimedia Tools Appl. **75**(6), 3221–3233 (2016)
22. Xie, S., Zhang, W., Ying, W., Zakim, K.: Fast detecting moving objects in moving background using ORB feature matching. In: 2013 Fourth International Conference on Intelligent Control and Information Processing (ICICIP), pp. 304–309. IEEE (2013)
23. Rublee, E., Rabaud, V., Konolige, K., Bradski, G.: ORB: an efficient alternative to SIFT or SURF. In: ICCV (2011)
24. Aguilar, W.G., Casaliglla, V.P., Pólit, J.L.: Obstacle avoidance based-visual navigation for micro aerial vehicles. Electronics **6**, 10 (2017)
25. Harris, C., Stephens, M.: A combined corner and edge detector. In: Alvey Vision Conference, pp. 147–151 (1988)
26. Torr, P., Zisserman, A.: MLESAC: a new robust estimator with application to estimating image geometry. Comput. Vis. Image Underst. **78**, 138–156 (2000)

# Machine Learning

# Cyberbullying Classification Using Extreme Learning Machine Applied to Portuguese Language

Jim Jones da Silveira Marciano[1(✉)] ⓘ,
Eduardo Mazoni Andrade Marçal Mendes[2] ⓘ,
and Márcio Falcão Santos Barroso[3] ⓘ

[1] College Presidente Antônio Carlos, Rod MG 482, s/n, Gigante,
Conselheiro Lafaiete, Minas Gerais 36400-000, Brazil
jimjonessm@gmail.com
[2] Federal University of Minas Gerais, Av. Antônio Carlos 6627,
Pampulha, Belo Horizonte, Minas Gerais 31270-901, Brazil
emmendes@cpdee.ufmg.br
[3] Federal University of São João del-Rei, Praça Frei Orlando 170, Centro,
São João del-Rei, Minas Gerais 36307-352, Brazil
barroso@ufsj.edu.br

**Abstract.** Increasing accessibility to virtual environments has resulted in higher incidences of Cyberbullying attacks and the consequences of these attacks can affect the victim's life for a long time or even permanently. For this reason, it is extremely important to develop tools to inhibit such practices. In this paper, a solution to the Cyberbullying classification problem using machine learning (Extreme Learning Machine) is proposed. The application of the proposed method to a database of sentences in Portuguese language shows promising results where compared to a standard method available in the literature.

**Keywords:** Machine learning · Cyberbullying · Extreme Learning Machine · Natural Language Processing · Artificial Neural Network

## 1 Introduction

With the result of social and technological developments, many social problems have also evolved together, becoming more visible and comprehensive. This happens with the *Bullying,* which is one of the greatest social problems for teenagers and young adults [1–3].

The bullying is to act in hostile, aggressive (verbally or physically) towards a person or group, consciously aimed at the embarrassment and humiliation of the target person [4]. There are several reasons for the occurrence of Bullying and research works have showed some reasons for it such as personality disorder, jealousy, religious and cultural differences, gender, among other factors [5, 6].

As the environment of living of people has been expanded to a digital universe, automatically Bullying migrated to it and became what is called the Cyberbullying. Cyberbullying is defined as an adaptation of bullying practices for digital media

D.A.C. Barone et al. (Eds.): LAWCN 2017, CCIS 720, pp. 109–117, 2017.
https://doi.org/10.1007/978-3-319-71011-2_9

[3, 7, 8]. The Cyberbullying practice exhibit some characteristics that enables the aggressor a wide range of attack that goes from the number of victims and to the visibility of his/her attitude. According to Donnerstein [9], many of the responses submitted by a person who is the target of Cyberbullying are exactly the same reactions submitted by victims of traditional bullying, demonstrating that the relationship between action or result remains the same.

The Bullying aggressor searches on the victims characteristics that can affect or scare the person in question. These features can be physical (some kind of disability, physical shape, skin color, hair type, etc.) or psychological (excessive shyness, intelligence, attention deficit disorder, etc.) [10].

In Neto's work [10], some interesting data is given to show the motivational context of this research. In the work, experiments were carried out to establish the relationship of people suffering Bulling in Brazil and the obtained results are shown below:

- 40.5% of people admitted to be directly connected to bullying;
- 60.2% stated that bullying occurs more frequently in the classroom;
- 80% had negative feelings towards acts of bullying, such as fear, grief and sorrow;
- 41.6% of those who admitted to being the target of bullying said they had asked for help of colleagues, teachers or family;
- 23.7% of those who asked for help received it;
- 69.3% acknowledged not knowing the reasons that cause the occurrence of bullying;
- 51.8% of Bullying aggressors said they received no guidance or warned about the severity of their actions.

To develop a tool for aggression identification of Cyberbullying it is important to understand the context of general bullying and Cyberbullying and to identify behavior features of individuals involved in it [10–12]. It is also extremely important to understand the language in which the textual Cyberbullying occurs, since, in the current scenario of digital social media, there is a very large variation in the written form, making difficult the development of a tool capable of distinguish an aggression from a joke. Such confusion comes from the fact the written form of the two situations can be the same in most practical situations.

As Cyberbullying has become a worldwide epidemic, better understanding of it has been receiving great attention lately as well the development of tools to identify this type of aggression [13–17]. Following this trend on the search of a tool to classify aggression of Cyberbullying, this paper aims to present a measure of Cyberbullying based on machine learning by proposing an Extreme Learning Machine classifier.

## 2 Extreme Learning Machine

The Extreme Learning Machine (ELM) is a tool that is part of the family of Artificial Neural Networks. The Artificial Neural Networks are sets of processing units that are capable of acquiring, maintaining and processing information [18]. Each processing unit is called neuron, since the structure of the processing is vaguely based on the network structure of the human nervous system.

Mcculloch was the author of the first publication on Artificial Neural Networks [19], where he developed a mathematical model inspired by the biological neuron. The first training method was proposed in 1949 and it was called learning rule of Hebb [20]. Since then, a vast number of papers have been published on the solution of a multitude of problems using neural networks.

The ELM (Fig. 1) is a feedforward network that emerged from a single hidden layer (single-hidden layer Feedforward Neural Network - SLFN). The ELM can be considered as the evolution of the SLFN training process with random nodes.

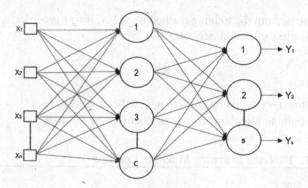

**Fig. 1.** Representation of an Extreme Learning Machine: ($n$) is the number of input data, (C) the number of neurons in the hidden layer and ($s$) the number of neurons in the output layer.

SLFNs uses $(x_i, y_i)$ data, where $x_i = [x_{i,1}, x_{i,2} \ldots x_{i,n}]^T$ is the vector of inputs data and $y_i = [y_{i,1}, y_{i,2} \ldots y_{i,s}]^T$ is the vector of outputs data. $n$ is the amount of nodes in the input layer and $s$ the amount of nodes in the output layer. A SLFN can be modeled according to Eq. (1) [21]:

$$\sum_{i=0}^{C} \beta_i g(x_j) = \sum_{i=0}^{C} \beta_i g(w_i \cdot x_j + b_i) = o_j,$$

$$for \quad j = 1 \ldots n$$

(1)

In Eq. (1), $n$ represents the quantity of samples in $x$ and $C$ represents the quantity of nodes at the hidden layers. In Eq. (1) there is also the weight vector for the input layer $w_i$; the weight vector for the hidden layer $\beta_i$; and the activation function $g(\cdot)$; and $b_i$ is the bias of the ith hidden node. In an easy way to understanding the training of a SLFN is to consider the equation (Eq. 2):

$$H \beta = Y,$$

(2)

where $H$ is the hidden layer output matrix (Eq. 3), $\beta$ (Eq. 4) the weights of hidden layer and $Y$ (Eq. 4) the respective outputs.

$$H(w,b,x) = \begin{bmatrix} g(w_1 \cdot x_1 + b_1) & \cdots & g(w_C \cdot x_1 + b_C) \\ \vdots & \cdots & \vdots \\ g(w_1 \cdot x_N + b_1) & \cdots & g(w_C \cdot x_N + b_C) \end{bmatrix}_{N \times C} \tag{3}$$

$$\beta = \begin{bmatrix} \beta_1^T \\ \vdots \\ \beta_C^T \end{bmatrix}_{C \times s} \quad e \quad Y = \begin{bmatrix} Y_1^T \\ \vdots \\ Y_N^T \end{bmatrix}_{N \times s} \tag{4}$$

In [21], starting from the training method of SLFN, it was proposed a new training strategy for this class of neural networks (Eq. 5):

$$\hat{\beta} = H^\dagger Y, \tag{5}$$

where $H^\dagger$ is the pseudoinverse of $H$, it can be rewritten as $H^\dagger = (H^T H)^{-1} H^T$. Therefore, the training algorithm can be summarized as follows:

---

**Algorithm 1:  Extreme Learning Machine Training**

---

```
   Entry: X, Y, C
1 Begin
2     for i = 1 to C do
3         Get wᵢ and bᵢ randomly
4     end
5     Get the output of the hidden layer, H matrix.
6     Calculate the outputs weights (β).(Equation 5)
7 end
```

---

## 3  Methodology

Python was the programming language used in this work along with the toolkit called Natural Language Toolkit (NLTK) [22] that has important tools for Natural Language Processing. Such a choice provided the necessary framework to quickly develop the basic structures of the work.

In this framework, the Floresta Corpus was chosen for processing Portuguese language in order to get the parsing of sentences to be analyzed. It was also necessary to obtain an auxiliary dictionary for the new words and for abbreviations, since some sentences did not fit the formal way of writing.

The phrases used in this work were obtained from social networks (Facebook, Twitter), Brazilian news sites (G1, UOL) and from a collaborative tool developed that people could give their contributions with phrases typical of Cyberbullying aggression. In the end, 153 sentences of aggression with the answers of the victims and 160 messages that was not of Cyberbullying aggression were collected and saved.

For the implementation of training, the words in a sentence were separated in pronouns, substantive, adjectives and verbs together with their frequencies of occurrences in Cyberbullying attacks. When one or more words of the selected category show up, the codes of the words and their frequencies are added, resulting always in an input vector *x* of 16 positions. When the attack does not receive an answer back, the values of the code of the words are set to zero and their frequency is the frequency of Cyberbullying phrase attack without answer. Algorithm 2 shows the structure of ELM training.

**Algorithm 2: Training and Validation Algorithm**

```
  Entry: phrases
1 begin
2     for i=1 to length(phrases) do
3        Get the syntactic parsing of the phrase [i]
4        Get the syntactic parsing of the answer[i]
5        Get the Cyberbullying frequencies of the pron.,
subst., adj. and verbs
6        Increasing the data obtained at input data matrix
X
7        Increase at output vector Y the outputs of the
data obtained
8     end
9     Randomize the positions of the X matrix and Y vec-
tor
10    Divide the training data and the validation data
11    Run the ELM training function
12    Run the ELM validation function
13 end
```

However, the final structure of the Cyberbullying classifier has the tasks specified in the following process diagram (Fig. 2).

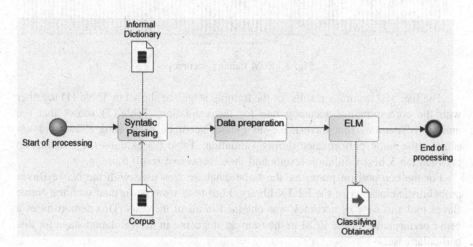

**Fig. 2.** Cyberbullying classifier process diagram.

## 4  Results

All sentences collected c tools were stored in a database that also contains the words of the formal dictionary and some informal words of Portuguese language. Every time a word shows up in one attack context, the Cyberbullying frequency of these words is incremented.

Algorithms 1 and 2 showed in Fig. (2) were implemented in Python. To obtain the results the following steps were considered:

- Interval of the number of neurons in the hidden layer from 1 to 100.
- For each variation of the amount of the neurons in the hidden layer the Algorithm 2 was executed 4 times.

Figure 3 shows the accuracies obtained on the training steps for the interval of the number of neurons at the hidden layer of the ELM, and Fig. 4 shows the accuracies obtained during the network validation. The best results were those were four executions for each set of neurons were performed.

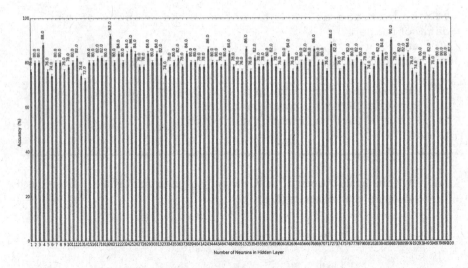

**Fig. 3.** ELM training accuracy.

The five best accuracy results for the training stage are shown in Table (1) together with the corresponding accuracy rate for the validation. Table (2) shows that the number of neurons that have achieved greater accuracy in training (92%) did not achieve the same performance during validation. Table (2) exhibits the relationship between the 5 best validation results and their respective result training.

For the comparison purposes, the same database was used with the Naive Bayes probabilistic classifier of the NLTK library. Four tests were performed with the Naive Bayes tool and 61% of accuracy was obtained in all of the tests. This demonstrates a better performance of the ELM in the training stage and in the validation stage for this particular case.

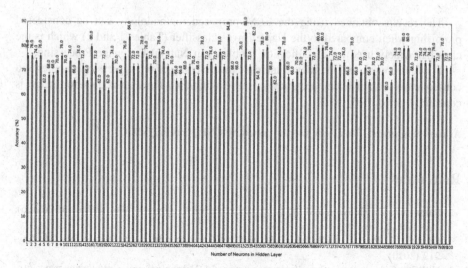

**Fig. 4.** ELM validation accuracy.

**Table 1.** Relationship between the training and validation accuracy.

| Neurons | Training % | Validation % |
|---------|-----------|--------------|
| 20      | 92        | 72           |
| 86      | 90        | 66           |
| 4       | 88        | 76           |
| 72      | 88        | 74           |
| 24      | 86        | 76           |

**Table 2.** Relationship between the training and validation accuracy.

| Neurons | Validation % | Training % |
|---------|--------------|------------|
| 51      | 86           | 86         |
| 25      | 84           | 86         |
| 48      | 84           | 84         |
| 54      | 82           | 80         |
| 16      | 80           | 80         |

## 5   Conclusions

In this paper, the classification of Cyberbullying using ELM as a learning mechanism was introduced and analyzed.

The widespread news about the effects on the people in general due to the Cyberbullying emphasizes the importance of the development of tools to mitigate or to eliminate such a problem not only in Brazil but around the world, since the victims of these attacks may carry the consequences throughout their lives.

The results obtained using the ELM proposed methodology can be considered promising when compared to the Naive Bayes classifier (Tables 1 and 2) which is the classifier widely used for classification problems in Natural Language Processing.

For future work and in order to increase the diversity of the aggression samples and consequently obtain better accuracy in the evaluations, the data acquisition stage is constantly performed.

**Acknowledgments.** This work was supported by CNPq and FAPEMIG.

# References

1. Salmivalli, C.: Bullying and the peer group: a review. Aggress. Viol. Behav. **15**(Março), 112–120 (2010)
2. Hinduja, S., Patchin, J.: Bullying, cyberbullying, and suicide. Arch. Suicide Res. **14**, 206–212 (2010)
3. NSPCC: Cyberbullying. http://www.nspcc.org.uk/. Disponível em: <http://www.nspcc.org.uk/help-and-advice/for-parents/online-safety/cyberbullying/cyberbullying_wda99645.html>. Accessed 29 May 2014
4. WWW.BULLYING.ORG: www.bullying.org. s.d. Disponível em: <http://www.bullying.org/external/documents/Bullying_Myths-Facts_Portuguese.pdf>. Accessed 07 sept 2011
5. Stockdale, L.A., et al.: Borderline personality disorder features, jealousy, and cyberbullying in adolescence. Personal. Individ. Differ. **83**, 148–153 (2015)
6. Tanrikulu, I., Campbell, M.: Correlates of traditional bullying and cyberbullying perpetration among Australian students. Child Youth Serv. Rev. **55**, 138–146 (2015)
7. Santomauro, B.: Cyberbullying: a violência virtual. NOVA ESCOLA, no. 223 (2010)
8. NOBULLYING.COM: Cyberbullying and bullying statistics 2014, finally! NOBullying.com. Disponível em: <http://nobullying.com/cyberbullying-bullying-statistics-2014-finally/>. Accessed 21 May 2014
9. Donnerstein, E.: Internet Bullying. Pediatr. Clin. North Am. **59**, 623–633 (2012)
10. Neto, A.A.L.: Bullying comportamento agressivo entre estudantes. Jornal de Pediatria **81**, S164–S172 (2005)
11. Wendt, G.W., Lisboa, C.S.M.: Agressão entre pares no espaço virtual: definições, impactos e desafios do cyberbullying. Psicologia Clinínica **25**, 73–87 (2013)
12. Tomşa, R., et al.: Student's experiences with traditional bullying and cyberbullying: findings from a Romanian sample. Procedia – Soc. Behav. Sci. **78**(Maio), 586–590 (2013)
13. Lieberman, H., Dinaka, K., Jones, B.: Let's gang up on cyberbullying. IEEE Computer Society (2011)
14. Almeida, R.J.A.: ESTUDO DA OCORRÊNCIA DE CYBERBULLYING CONTRA PROFESSORES NA REDE SOCIAL TWITTER POR MEIO DE UM ALGORITMO DE CLASSIFICAÇÃO BAYESIANO. Linguagem e Tecnologia, Texto Livre (2012)
15. Moore, M.J., et al.: Anonymity and roles associated with aggressive posts in an online forum. Comput. Hum. Behav. **28**, 861–867 (2012)
16. Nocentini, A., Zambuto, V., Menesini, E.: Anti-bullying programs and information and communication technologies (ICTs): a systematic review. Aggress. Viol. Behav. **23**, 52–60 (2015)
17. Smith, D.A., Lieberman, H.: Helping users understand and recover from interpretation failures in natural language interfaces. ACM (2012)

18. Haykin, S.: Neural Networks: A Comprehensive Foudation, 2nd edn. Prentice Hall, Upper Saddler River (1999)
19. McCulloch, W.S., Pitts, W.: A logical calculus of the ideas immanent in nervous activity. Bull. Math. Biophys. **5**, 114–133 (1943)
20. Hebb, D.O.: The Organization of Behavior: A Neuropsychological Theory. Wiley, New York (1949)
21. Huang, G.-B., Zhu, Q.-Y., Siew, C.-K.: Extreme learning machine: a new learning. In: International Joint Conference on Neural Networks, pp. 985–990 (2004)
22. Bird, S., Klein, E., Loper, E.: Natural Language Processing with Python: Analyzing Text with the Natural Language Toolkit. O'Reilly Media, Sebastopol (2009)

# Pseudorehearsal Approach for Incremental Learning of Deep Convolutional Neural Networks

Diego Mellado, Carolina Saavedra[(⊠)], Stéren Chabert, and Rodrigo Salas

Engineering Faculty, Biomedical Engineering School, Universidad de Valparaíso, General Cruz 222, Valparaíso, Chile
diego.mellado@postgrado.uv.cl, {carolina.saavedra,rodrigo.salas}@uv.cl

**Abstract.** Deep Convolutional Neural Networks, like most connectionist models, suffers from catastrophic forgetting while training for a new, unknown task. One of the simplest solutions to this issue is adding samples of previous data, with the drawback of increasingly having to store training data; or generating patterns that evoke similar responses of the previous task.

We propose a model using a Recurrent Neural Network-based image generator in order to provide a Deep Convolutional Network a limited number of samples for new training data. Simulation results shows that our proposal is able to retain previous knowledge whenever some few pseudo-samples of previously recorded patterns are generated.

Despite having lower performance than giving the network samples of the real dataset, this model is more biologically plausible and might help to reduce the need of storing previously trained data on bigger-scale classification classification models.

**Keywords:** Deep learning · Incremental learning · Computational neuroscience · Stability Plasticity dilemma

## 1 Introduction

Deep learning has become on recent years one of the main buzzwords on the artificial neural network field, due to the increased access to high-processing equipment and increasing number of applications in vision research, natural-language processing, among others. This concept refers to the use of multiple

---

D. Mellado—The authors acknowledge the support by Chilean Grants "Proyecto estudiantes de los convenios de desempeño UVA 1315, UVA 1401 and UVA 1402" from Universidad de Valparaíso, and CONICYT + PAI/CONCURSO NACIONAL INSERCIÓN EN LA ACADEMIA, CONVOCATORIA 2014 + Folio (79140057). The work of R. Salas was partially funded by project grant FONDEF IDEA ID16I10322.

© Springer International Publishing AG 2017
D.A.C. Barone et al. (Eds.): LAWCN 2017, CCIS 720, pp. 118–126, 2017.
https://doi.org/10.1007/978-3-319-71011-2_10

models of artificial neural networks in conjunction, forming a multi-layered network, capable of doing different interpretations on complex data. These models have the particularity of learning how different input features relate to each other, learning representations by abstraction, as done in the human brain [1]. But despite of its multiple advantages over numerous Machine Learning algorithms, Deep learning models still share some of their common difficulties for learning new, unknown data for the network.

Several simulation results show that when static models have sequential learning in non-stationary environments, they catastrophically forget the previously learned patterns [2,3]. Different solutions to this problem have been proposed on different neural network models, based on our current understanding on how biological brains understand newer knowledge, and apply and integrate to previous neural pathways. For example, by modifying the plasticity of a network's weights while training a new task [4].

Previous work by the authors has shown how Catastrophic Forgetting impacts on neural network models, and how rehearsal and pseudo-rehearsal approaches can help these to retain knowledge on incremental training [5,6].

In this paper we propose a model to overcome catastrophic forgetting by automatically creating pseudo-samples from stochastic information of already learned patterns. In our solution, the previously learned data are non-longer needed. Since we generate random patterns capable of evoking old knowledge while training new tasks. This allows us to dispose already learned data for training, effectively reducing the memory cost of storage of old data.

The remainder of this paper is organized as follows. In Sect. 2 we introduce the concept of catastrophic forgetting and some of the present solutions to this problem. Section 3 explains how we apply these solutions on a deep convolutional network. In Sect. 4, we present the results of our implementation. Finally, in Sect. 5 we discuss on how our model can be improved.

## 2    Catastrophic Forgetting

For most machine learning algorithms, one of the main problems while training new data is the presence of new, previously unknown features, not present on the original training. This phenomenon affects the learning process on most connectionist network models [7,8], making them unable to retain old training tasks while learning new data, i.e. the new learned information most often erases the one previously learned.

One of the main factors that causes catastrophic forgetting is sequential learning, since the network overlaps representations made previously [9]. This effect is not cognitively plausible, in addition to the fact that it is disastrous for most practical applications, so efficient algorithms to deal with catastrophic forgetting are needed.

The problem is how to design a network that is simultaneously sensitive to a new input, but not disrupted by it. One challenge is to balance the ability of generalizing information from the presented data with the capacity of the network to retain the old information, for example using local representations [10]. This is known in neuroscience as the *"Stability versus Plasticity Dilemma"*.

This problem has been thoroughly studied on backpropagation-based networks [9,11], among other models. One solution for this problem is to select the rehearsal approach, consisting on picking a small number of items from the old training data, and combining it with the new untrained set in order to maintain the structure of the network. Two examples of these models are Ratcliff's recency rehearsal approach [7]. Using the last three previous items placed on a queue with the new item, effectively retraining them; and the addition of a random sample of previously trained data, known as random rehearsal [12].

The other solution is the pseudorehearsal approach, defined as the usage of populations of "pseudo-items" constructed from a random-input vector with a defined output, obtained from the already trained model, in order to maintain the structure of the network while training with new data [12]. One example of pseudorehearsal on a multilayer neural network is the work of Hattori and Hosaka [13], using this model with a *no-propagation* algorithm-based network. Their main difference with other pseudorehearsal models comes with the usage of subsets of pseudopatterns in proportion of the capacity of their training method.

## 3    Proposed Model and Method

We designed a deep neural network model combining different networks in order to overcome catastrophic forgetting. Our method consist of two stages: The first one is to train the model with an initial dataset, and the second stage is to train the model with unseen data, plus an small subset of evoked patterns from the previous stage.

We evaluated our model using the MNIST database [14], a classical image classification dataset of handwritten numbers from 0 to 9. The first task for our algorithm was to learn and correctly classify images from classes $\{0, \ldots, 4\}$ and the second task was to learn the remaining classes $\{5, \ldots, 9\}$, retaining the previous knowledge. An example of the proposed model architecture, is shown on Fig. 1. In this example, the model was previously trained using the first set (i.e. the number 3). On the second stage, the number 8 will be teached. In order to achieve this purpose, the number 3 is evoked using an image generator, explained in the following section. As a result, the model is able to learn new patterns and retain the older ones.

### 3.1    Image Generator Network (IGN)

At first, we implemented an image generation network based on DRAW, a Recurrent Neural Network ($RNN$) structure commonly used for image generation [15] on its non-selective attention form.

Let $RNN^{dec}$ be the function enacted by the decoder network at a single time-step. The output of $RNN^{dec}$ at time $t$ is the encoder hidden vector $h_t^{dec}$. Correspondingly, the output of the encoder $RNN^{enc}$ is denoted by $h_t^{enc}$. At

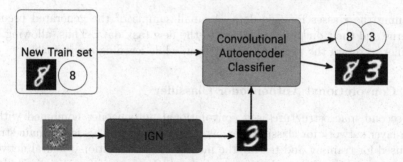

**Fig. 1.** Proposed model architecture, training a new dataset with the addition of random patterns from the previous training.

the first time-step the variables $c_0$, $h_0^{enc}$, $h_0^{dec}$ are initialised to learned biases, for a sample image $x$ we compute the following operations for each time step $t = 0 \ldots, T$

$$\hat{x}_t = x - \mathcal{S}(c_{t-1})$$
$$r_t = [x, \hat{x}_t]$$
$$h_t^{enc} = RNN^{enc}\left(h_{t-1}^{enc}, [r_t, h_t^{dec}]\right)$$
$$z_t \sim Q(Z_t|h_t^{enc})$$
$$h_t^{dec} = RNN^{dec}\left(h_{t-1}, z_t\right)$$
$$c_t = c_{t-1} + W(h_t^{dec})$$

with $\mathcal{S}$ being the sigmoid function $\mathcal{S}(x) = \frac{1}{1+e^x}$, $Q$ being defined as a Gaussian distribution with parameters $(\mu, \sigma^2)$ obtained from $h_t^{enc}$, forming a *latent vector* defined as $z_t$ and $W(x)$ being a fully-connected layer. At the end of the temporal sequences from the RNN, this network outputs into a Sigmoid activation function, presenting as an output an image array. For our model, we implemented a simplified version of the network and compensated with modifying its depth, adding a second RNN layer on both encoding and decoding phases.

**Pseudo-Pattern Generation.** In order to train the network for a new task without using the original data from the previous learning task (as done on rehearsal models), the image generator creates pseudo-patterns based on samples obtained from a Normal distribution (1), using the mean output information from the neural network of a specific task $\bar{x}_y$.

$$\tilde{x} \sim \mathcal{N}\left(\mu(\bar{x}_y), 2\sigma(\bar{x}_y)\right) \tag{1}$$

with $\mu_{\bar{x}_y}$ being the mean of each point of $\bar{x}$, and $\sigma$ their standard deviation.

This allows the model to create new samples from random data, inspired from what it has learned. Those pseudo-patterns are then evaluated on the classifier, obtaining the classes for this data. A small sample of the generated pseudo-patterns and their outputs are added to the new task data. Thus, allowing the classifier to retain the previous information, while learning a new task.

## 3.2  Convolutional Autoencoder Classifier

The second macrostructure is a convolutional autoencoder combined with a multi-layer network for classification, as shown on Fig. 2. This is the main structure used for training and testing the model. The convolutional neural network consist on a series of Convolutional layers with Glorot-uniform initialized filters [16] of size $7 \times 7$, $5 \times 5$ on pairs of 2 layers and $3 \times 3$ on the final layer, using Rectifier Linear Unit (*ReLU*) activation for each layer's output; and MaxPooling layers in between. After these, the network splits in a mirrored version of these layers, effectively creating an autoencoder; and a Fully Connected classifier network, using two layers of densely connected neurons with Sigmoid activation, ending with a output layer of with Softmax activation, corresponding to the number of labels the model tries to learn. The autoencoder output corresponds to the model's interpretation from the input.

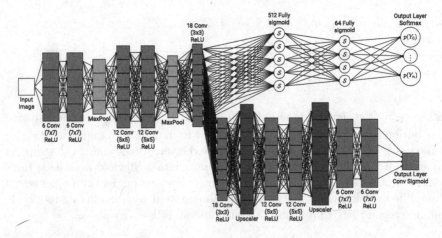

**Fig. 2.** Convolutional autoencoder neural network with classification stage.

The gradient loss for each structure was calculated using binary cross-entropy for the image generator and the autoencoder, while the classifier's loss was calculated using categorical cross-entropy (2).

$$H(p,q) = -\sum_i p_i \log(q_i) = -y \log(\hat{y}) - (1-y) \log(1-\hat{y}) \qquad (2)$$

where $p, q$ are the probabilities for the real $(y)$ and predicted $(\hat{y})$ outputs respectively, $p_{y=1} = y$, $p_{y=0} = 1 - y$ and $q_{y=1} = \hat{y}$ and $q_{y=0} = 1 - \hat{y}$ for each class $i$.

## 3.3 Experiment

As explained before, the initial stage was to train the model with the initial dataset. In order to prepare for the second stage, the image generator is used to generate 512 initial pseudo-samples from each class and then tested on the classifier in order to guarantee that these are identified by the network. From this set, we only kept the generated images that the classifier had a 95% of confidence of being an identified number, and the others were discarded from the second training task. From those, we added an small sample to the second stage dataset, starting from 5 up to 300 patterns of each class from the previous stage. After the second stage, the classifier was evaluated using the unseen test dataset, composed by all numbers from 0 to 9, measuring its performance using the prediction accuracy. In order to evaluate our algorithm, we compared our method with a classical rehearsal implementation of our classifier, measuring its training accuracy when added samples of previous training images.

For the image generator, the network was trained for 40 epochs in batches of 64 samples, while the classifier was trained for 5 epochs in batches of 32 samples on each training stage. The model was programmed using Keras [17] on Python 3.6 and run on a HP Proliant ML110 with a NVIDIA GeForce GTX980 GPU for processing.

## 4 Results

Our model when trained with the complete dataset, has a mean accuracy of $98.98 \pm 0.14\%$ and this is considered our peak performance of our setup. For our experiment, we measured the training loss and performance of our image generator using mean squared error (MSE) for each training epoch, and the training loss of the classifier on both training tasks, as shown on Fig. 3.

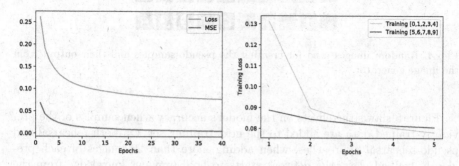

**Fig. 3.** Left: Training loss and MSE of our image generator implementation. Right: Training loss at both steps of the classifier structure.

After training the model for the first training task, we generated the pseudo-samples using the method presented in (1), obtaining the samples shown on

Fig. 4. From all 512 images for each class, the classifier correctly identified (defined as having a 95% of confidence of being that number) on around $88.15 \pm 3.67\%$ of them on each training experiment. Table 1 shows how each input from the generated data was distributed on each class. Class 3 was identified the most, adding samples from other classes; while class 1 was the least correctly identified from the whole dataset. This is due to broken digits formed, making them too thin to be recognized or mistaken for another number. Nevertheless, on every but one experiment, the number of pseudo-samples accepted were the same for each class.

**Table 1.** Pseudo-samples correctly identified as input for classifier

| Classes | N° of images | % of original |
|---|---|---|
| 0 | $480.75 \pm 34.79$ | $91.94 \pm 6.80$ |
| 1 | $308.81 \pm 89.13$ | $60.31 \pm 17.41$ |
| 2 | $476.00 \pm 40.17$ | $92.97 \pm 7.85$ |
| 3 | $583.56 \pm 45.15$ | $113.98 \pm 8.82$ |
| 4 | $417.50 \pm 37.82$ | $81.54 \pm 7.39$ |

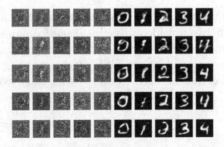

**Fig. 4.** Random images used for creating the pseudo-samples and their output from the image generator.

Figure 5 shows the effects on the model's accuracy when samples of the previously trained data are added to the next training set. On both rehearsal and pseudo-rehearsal approaches, when adding more than 20 images of each previously trained class, the network starts to hold previous knowledge from the earlier task. And with more than 50 samples per class, both systems start reaching their top performance. The increasing performance of both models can be modelled with a logarithmic function.

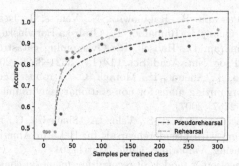

**Fig. 5.** Model accuracy after incremental training using rehearsal and pseudo-rehearsal approaches

# 5  Discussion and Future Work

The proposed model performed as expected when using random patterns generated from the previous training, making it feasible to train in increasing steps with newer classes without having to store older data for training.

Although the rehearsal method reaches significantly higher performance compared to a pseudo-rehearsal approach, this is not biologically viable because memory doesn't retain the exact representations of previous learning. And computationally when training with increasingly bigger datasets, storage can be problematic in order to constantly store a training subset.

Thus, creating pseudo-samples using an approximation of each class (i.e. using the mean and standard deviation of data) is more plausible and less memory-intensive. Saving these parameters after training a task and using them to create pseudo-samples available at any time can save memory, compared to storing part of the previous training set. This could improve storage issues of previously trained data, specially when adding newer tasks at hand.

Further improvements can be done to the model, specially on the image generator implementation in order to make it more robust and capable of effectively creating patterns recognizable by the network with few parameters. The model can also be improved by making the network capable of recognizing without input the appearance of unknown tasks and thus, able to adapt and grow if needed.

# References

1. Dhar, V.: The scope and challenges for deep learning. Big Data **3**(3), 127–129 (2015)
2. French, R.M.: Catastrophic forgetting in connectionist networks. Trends Cogn. Sci. **3**(4), 128–135 (1999)
3. Crossberg, S.: Studies of Mind and Brain: Neural Principles of Learning, Perception Development, Cognition and Motor Control. D. Reidel Publishing, Dordrecht (1982)

4. Kirkpatrick, J., Pascanu, R., Rabinowitz, N., Veness, J., Desjardins, G., Rusu, A.A., Milan, K., Quan, J., Ramalho, T., Grabska-Barwinska, A., Hassabis, D., Clopath, C., Kumaran, D., Hadsell, R.: Overcoming catastrophic forgetting in neural networks. Proc. Nat. Acad. Sci. **114**(13), 3521–3526 (2017)
5. Salas, R., Moreno, S., Allende, H., Moraga, C.: A robust and flexible model of hierarchical self-organizing maps for non-stationary environments. Neurocomputing **70**(16), 2744–2757 (2007)
6. Mellado, D., Salas, R., Chabert, S., Veloz, A., Saavedra, C.: Incremental ensemble of deep learning with pseudorehearsal. In: II Latin American Conference on Statistical Computing, Valparaiso, Chile, March 2017
7. Ratcliff, R.: Connectionist models of recognition memory: constraints imposed by learning and forgetting functions. Psychol. Rev. **97**(2), 285–308 (1990)
8. McCloskey, M., Cohen, N.J.: Catastrophic interference in connectionist networks: the sequential learning problem. In: Bower, G.H. (ed.) Psychology of Learning and Motivation, vol. 24, pp. 109–165. Academic Press, New York (1989)
9. French, R.M.: Pseudo-recurrent connectionist networks: an approach to the 'sensitivity-stability' dilemma. Connect. Sci. **9**(4), 353–380 (1997)
10. French, R.M.: Semi-distributed representations and catastrophic forgetting in connectionist networks. Connect. Sci. **4**(3–4), 365–377 (1992)
11. Robins, A., McCallum, S.: Catastrophic forgetting and the pseudorehearsal solution in hopfield-type networks. Connect. Sci. **10**(2), 121–135 (1998)
12. Robins, A.: Catastrophic forgetting, rehearsal and pseudorehearsal. Connect. Sci. **7**(2), 123–146 (1995)
13. Hattori, M., Hosaka, T.: Applying pseudorehearsal to multilayer neural networks trained by no-prop algorithm. In: Advances in Computational Intelligence, November 2015
14. LeCun, Y., Bottou, L., Bengio, Y., Haffner, P.: Gradient-based learning applied to document recognition. Proc. IEEE **86**(11), 2278–2324 (1998)
15. Gregor, K., Danihelka, I., Graves, A., Rezende, D.J., Wierstra, D.: DRAW: a recurrent neural network for image generation, February 2015. arXiv: 1502.04623 [cs]
16. Glorot, X., Bengio, Y.: Understanding the difficulty of training deep feedforward neural networks. In: Proceedings of 13th International Conference on Artificial Intelligence and Statistics, pp. 249–256 (2010)
17. Chollet, F., et al.: Keras (2015). https://github.com/fchollet/keras

# Graphics Systems and Interfaces

# A Brain Computer Interface Using VEP and MMSC for Driving a Mechanical Arm

Marcos Antônio Abdalla Júnior[1]([✉]), Carlos Alberto Cimini Júnior[1],
Márcio Falcão Santos Barroso[2], and Leonardo Bonato Félix[3]

[1] Universidade Federal de Minas Gerais, Belo Horizonte, Brazil
marcosabdalla@yahoo.com.br, carlos.cimini@gmail.com
[2] Universidade Federal de São João del Rei, São João del-Rei, Brazil
barroso@yfsj.edu.br
[3] Universidade Federal de Viçosa, Viçosa, Brazil
leobonato@ufv.br

**Abstract.** The following article presents the development of a BCI system using Visual Evoked Potential and a detection system based on Multiple Mean Squared Coherence method. The developed BCI was used to control a small robotic arm.

**Keywords:** Visual Evoked Potential · Brain Computer Interface · Multiple coherence

## 1 Introduction

The term assistive technologies (AT) is used to describe the assets that takes in account, all the interdisciplinary characteristics of knowledge, products, resources, methodologies, strategies, services and practices for promoting functionalities relative the participation of deficient, incapable and reduced mobility's personal for more autonomy, independence, life qualities and social inclusion. When coupled with a brain computer interface (BCI), this new system can provide a non-muscular way of communication for people, e.g. those who suffer from severe palsy, brain stem stroke or Amyotrophic Lateral Sclerosis (ALS). This system can establish a direct connection between the human brain and a computer [1]. The BCI system is able to acquire the electroencephalographic (EEG) from the brain, interpret and provide a provide a neuroprothesis control. control. Surface electrodes placed in the scalp do this.

One of the main reasons for people with ALS to use a BCI system is that these patients loose the abilities to initiate and control all voluntary movement, although eye movement are not always, but usually spared until the terminal stages of the disease [2].

There are several common ways to obtain the brain's response, Movement-related Potential (MRPs) and other sensorimotor activities [3], Slow Cortical Potentials (SCPs) [1], P300 [4], Response to mental tasks [5]. Activity of Neural Cells (ANC) [6], Visual Evoked Potentials [7]. A well-known technique of visually evoked potential is the Steady State Visually Evoked Potential or SSVEP. The SSVEP consists in producing a stimulating light that flickers at a fix frequency. This stimulation is often done using:

© Springer International Publishing AG 2017
D.A.C. Barone et al. (Eds.): LAWCN 2017, CCIS 720, pp. 129–142, 2017.
https://doi.org/10.1007/978-3-319-71011-2_11

images from television sets or computer monitors, Xe-light, fluorescent light and light-emitting diode (LED) [8].

The interpretation of the SSVEP signals requires an Objective Response Detection (ORD) system. The ORD provides the detection of a stimuli signal inside the EEG signal. An example of ORD is the Multiple Mean Squared Coherence (MMSC) proposed by de Sá et al. in [9]. The MMSC can detect a periodic signal immersed in a random signal. With this multiple system a large number, signals are acquired from different locals of the scalp.

As the SSVET uses the signals from the vision processing area of the brain, we the scalp electrodes of occipital area were used. In the work of Morgan and colleagues, the authors used the C3, C4, P3, P4, O1, O2, T5 and T6 areas of the brain. In the work of de Sa et al. [10] and in the work of Muller-Putz and Pfurtscheller [11] the authors chosen to use only the O1 and O2 channels.

This paper proposes the developing of a BCI based on the SSVEP and MMSC for driving a small robotic arm.

## 2 Materials and Methods

### 2.1 BCI, EEG History, Procedure and Recordings

Electrical potentials from brain activity can be measured on the surface of the cerebral cortex or the scalp, through electrodes. The Brain Computer Interface (BCI) is an approach that aims on translating these potentials into output signals that represent the user's intent to communicate with the environment without using peripheral nerves or muscles [7, 12, 13]. It is estimated that in the US alone, there are over 200.000 patients with spinal cord injuries that lead to motor disabilities. At least half of these patients are paralyzed from the neck down. There are also 5.000.000 survivors of heart attack and 400.000 amputees [14]. In these cases and in cases where the patients suffer of Amyotrophic Lateral Sclerosis, Seizures, damages in the brain or spinal cord, cerebral paralysis, muscular dystrophy or any other illness that can affect the path ways from the brain to the muscles [15–17]. The BCI is proving it self to be a technological tool for the rehabilitation and reintroduction of these patients back to their family and work.

It has attributed to Richard Caton (1842–1926), the first works in relation to the measurement of the electrical activities in the brain. For his studies, Canton used a Galvanometer and a ray of light reflected in the mirror inside the device, directed to a large scale attached to the wall [18]. Caton developed his work on monkeys, cats, mice and dogs, but it was Hans Berger (1873–1941), who began working on human patients. In the works from 1926, Berger used a powerful Siemens galvanometer with a double core. With this instrument and non-polarized electrodes, he made the first recordings of human EEG on photo paper with a duration of 1 to 3 min.

The EEG signal is composed of several waves of different frequencies. Among these, there are the dominant frequencies that are related to factors such as alertness, age, and electrode position, use of medications, presence or absence of diseases. A normal adult, relaxed and with closed eyes, has a dominant frequency in the parietal and occipital lobes in the range of 8 Hz to 13 Hz, called alpha rhythm ($\alpha$). With eyes open, the rhythm of the EEG become less synchronized and the frequency increases to the range of 13 Hz to 30 Hz, called beta rhythm (β). During the sleep, two other rhythms arise, the delta rhythms (δ) in the range of 0.5 Hz to 4 Hz and the theta rhythms (θ) 4 Hz to 7 Hz. Observing the motor cortex region there is the presence of mu rhythm (μ) in the range of 8 Hz to 13 Hz with variations in amplitudes when the individual makes some movement. Above 30 Hz and up to the 100 Hz there is the gamma rhythm (γ) and is linked to motor and cognitive functions [18].

There are different types of EEG signals for different applications. Spontaneous Signs (SS), Motor Imagery (MI), Evoked Potentials (EP). Inside the EP class of EEG response there is the Steady-State Visually Evoked Potential (SSVEP). SSVEP is a Visual Evoked Potential (VEP) made by flashing lights in a steady-state rhythm [19]. By generating a steady-state response, the signal becomes more robust with respect to the contamination of the signal by artifacts (e.g. eye movement and blinking).

## 2.2  The BCI System

The proposed BCI system if shown in Fig. 1. In this case, a computer generates the signal with the desired stimulus frequency and the trigger signal. The signal is than sent to the LED's triggering. The evoked responses made by the intermittent brightness of the LED's is captured by a BrainNet EEG recorder and sent via network cable to a second computer where the data will be processed in Matlab. When the processing ends, the commands are sent to a LabVIEW interface program that drives the arm.

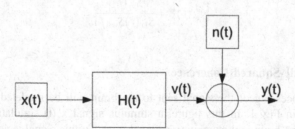

**Fig. 1.** Linear model to represent the relation of the stimulation signal and the cerebral response.

## 2.3  Coherence Function

Coherence is a mathematical tool widely used to quantify the similarity between signals such as EEG. Coherence differs from Correlation as it works with data in the frequency domain (as opposed to the correlation that works in time domain). Coherence is obtained by normalizing the cross-spectrum, which is the result of the division of the crossed spectrum by the product of the auto-spectrum Benignus (1969). For example, given the signals x (t) and y (t):

$$\gamma(f) = \frac{S_{xy}(f)}{\sqrt{S_{xx}(f)S_{yy}(f)}} \tag{1}$$

where $S_{xy}(f)$ is the crossed spectrum between the signal x (t) and y (t) given by:

$$S_{xy}(f) = \lim_{T \to \infty} \frac{1}{T} E[X^*(f,T)Y(f,T)], \tag{2}$$

where $X^*(f,T)$ and $Y(f,T)$ are respectively, the Fourier Transform of the signal x (t) and y (t) in stretches of T seconds, * denotes the complex conjugate, and T is the range of the signals under analysis. In the case where T is finite, the cross-spectrum estimate is:

$$\widehat{S_{xy}}(f) = \frac{1}{MT} \sum_{i=1}^{M} X_i^*(f,T)Y(f,T), \tag{3}$$

where (^) indicates estimate, M is the number of segments used in averaging. $S_{xx}(f)$ and $S_{yy}(f)$ are the auto spectrum given by:

$$\widehat{S_{xx}}(f) = \frac{1}{MT} \sum_{i=1}^{M} |x_i(f,T)|^2 \tag{4}$$

The squared module of Eq. 4 is called coherence and is restricted to a range from 0 to 1. The coherence function if given by:

$$\left| \gamma_{xy}^2(f) \right| = \frac{|S_{xy}(f)|^2}{S_{xx}(f)S_{yy}(f)} \tag{5}$$

## 2.4   Magnitude-Squared Coherence

At first, the processing of the signal sent to the brain will be modeled by the linear system shown in Fig. 1. In this figure, a stimulus signal x (t) simulating the image processing by the brain and generating a signal v (t). To this signal is added a noise n (t) not related to x (t) which models the background EEG signal. This sum generates the output y (t).

Dividing the discrete signal x[n] and y[n] (related according to Fig. 1) in M segments, the estimation of the coherence between these signals can be obtained by means of Eqs. 6 and 7 as fallows [10]:

$$\widehat{\gamma_{xy}^2}(f) = \frac{\left| \sum_{i=1}^{M} X_i^*(f)Y_i(f)^2 \right|}{\sum_{i=1}^{M} |X_i|^2 \sum_{i=1}^{M} |y_i|^2} \tag{6}$$

where $X_i(f)$ and $Y_i(f)$ are Fourier transforms of the ith window and M is the number of windows used in the estimation.

For a particular case where x[n] is a deterministic and periodic signal, $X_i(f)$ has the same value in each averaging window, then $X_i(f) = X(F)$, so the Eq. 6 can be simplified:

$$\widehat{\gamma_{xy}^2}(f) = \frac{\left|\sum_{i=1}^{M} Y_i(f)^2\right|}{M \sum_{i=1}^{M} |y_i|^2} \tag{7}$$

In order to distinguish from Eq. 6, the coherence between a random signal and a deterministic signal will be denoted by $\widehat{K_y^2}$:

$$\widehat{K_y^2} = \frac{\left|\sum_{i=1}^{M} Y_i(f)^2\right|}{M \sum_{i=1}^{M} |y_i|^2} \tag{8}$$

Thus, the coherence between a deterministic and periodic signal and a random signal can be estimated using only the random signal y[n] provided that the stimulus x[n] is guaranteed to be periodic.

## 2.5    Critical Values for $K_y^2(f)$

For a quantitative analysis of the EEG signal response, it is important to obtain values for which it can be said that there is no detected response (statistical threshold). Thus for the null hypothesis (H0) of absence of response, considering y[n] with Gaussian distribution, the sample distribution of $\widehat{K_y^2}$ is given by [10]:

$$(M-1)\frac{\widehat{K_y^2(f)}}{1 - \widehat{K_y^2(f)}} \sim F_{2,2(M-1)} \tag{9}$$

Where $\sim$ denotes "is distributed according to", $F_{2,2(M-1)}$ is the Fisher's F distribution with 2 and 2(M−1) degrees of freedom. By means of the Eq. 9, the critical values can be calculated as:

$$\widehat{K_y^2}crit = 1 - \alpha^{\frac{1}{M-1}} \tag{10}$$

where $\alpha$ is the significance level.

## 2.6    Extension of the MSC to the Multivariable Case (MMSC)

In this work, we will observe the spectral relation for a system with two simultaneous inputs that add up and to output according to Fig. 2.

In Fig. 2 x[n] is the signal formed from the stimulation of two different and simultaneous signals. This signal is filtered by Hi[n], representing the evoked response at different sites of the brain. The background EEG is represented by a noise ni[n] and the sum of vi[n], generates the collected response yi[n].

The estimate of the multiple coherence for the system of Fig. 2 is denoted by:

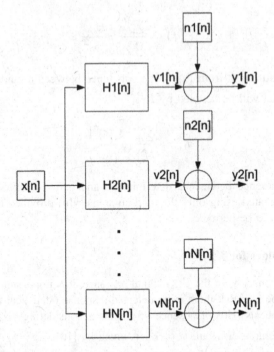

**Fig. 2.** Multivariable linear model to represent the relation of stimulation signal and brain responses.

$$\gamma^2_{x:y1,y2,\ldots,yn}(f) = \frac{\hat{S}^H_{yx}(f)\hat{S}^{-1}_{yy}(f)S_{yx}(f)}{\widehat{S_{xx}(f)}} \tag{11}$$

where $(\cdot)^H$ stands for the Hermitian of the matrix, f is the frequency and $\widehat{S_{xx}}(f)$ is the power spectrum estimate of x[n]. The cross-spectrum are calculated as:

$$\widehat{S_{ypyq}}(f) = \sum_{i=1}^{M} Y^*_{pi}(f)Y^*_{qi}(f) \tag{12}$$

where $Y^*_{pi}(f)$ is the Fourier transform of the i-th signal window $y_p[n]$. Similar as to the one variable case, if x[n] is a periodic and deterministic signal, and the windows are synchronized (each window must have the same number of oscillations), its Fourier transform will always have the same value in all the segments. Therefore, Eq. (11) can be written as the MMSC as:

$$\widehat{K^2_N}(f) = \frac{Y^H(f)S^{-1}_{yy}(f)Y(f)}{M} \tag{13}$$

where

$$Y(F) = \begin{bmatrix} \sum_{i=1}^{M} Y_{1i}^*(f) \\ \sum_{i=1}^{M} Y_{2i}^*(f) \\ \vdots \\ \sum_{i=1}^{M} Y_{Ni}^*(f) \end{bmatrix} \tag{14}$$

## 2.7   Critical Values for MMSC

Considering the linear model of Fig. 2 and taking into account that the number of output channels (N) is less than the number of windows (M) observed. According to [20] the theoretical critical values for $\widehat{K_N^2}(f)$ can be obtained as:

$$\widehat{K_{N\,crit}^2}(f) = \frac{F_{crit_{\alpha,2N,2(M-N)}}}{\left(\frac{M-N}{N}\right) + F_{crit_{\alpha,2N,2(M-N)}}} \tag{15}$$

## 2.8   Multiple Real-Time Coherence

Equation 13 presents the value of the multiple coherence used in off-line analyzes. However, in order to analyze the EEG response in real-time, the Eq. 13 had to be modified to a recursive form:

$$V(f,M) = \left[\sum_{i=1}^{M-1} Y_{1i}(f)\ \sum_{i=1}^{M-1} Y_{2i}(f) \cdots \sum_{i=1}^{M-1} Y_{Ni}(f)\right] + \left[Y_{1M}(f)Y_{2M}(f)\cdots Y_{NM}(f)\right]$$
$$\widehat{S_{ypyq}}(f,M) = \sum_{i=1}^{M-1} Y_{pi}^*(f)Y_{qi}(f) + Y_{pM}^*(f)Y_{qM}(f) \tag{16}$$

In this formulation, MMSC works with data from the current window (M) and from the past windows (M−1) as well as the values of $\widehat{S_{ypyq}}(f,M)$ also count as the values of the current windows and past.

## 2.9   The SSVEP Stimulators

An intermittent flashing light can generate high amplitude evoked potentials using low frequency bands (5–12 Hz), average frequencies (12–25 Hz) and high frequencies (25–50 Hz) [21]. With the use of Light-Emitting Diodes (LEDs), the evoked potentials can vary in the frequencies of 1–90 Hz [22]. In this project, we used the TELUXTM, model V LWW9900, white color, made by Vishay Semiconductors, with a total flow of 2200 mililumens, light intensity of 0.8 milicandela per mill lumens, TTL (Transistor-Transistor Logic) of 5 V. The current in the device was limited by means of a resistor of 33.00 ohms obtaining a current of 0.15 amps per LED.

The concentration commands, that is, where the individual should concentrate the vision, was done by means of a tricolor LED (red, green and blue) or RGB. For the commands, the PLCC − 6 LED from Avago Technologies was chosen. Using this mechanism, it was possible to give the user three different commands: when the red light is on, the user should not pay attention to any of the stimulators, in case the green or blue light turns on the user should focus on the left stimulator or right respectively.

## 2.10  Stimulation Frequency's

In this work, two frequencies of stimulation of the Theta band were adopted. Due to the use in several works [7, 23, 24], it was chosen the frequencies of 5 Hz and 7 Hz.

The signals evoked by means of intermittent stimulation have the same fundamental frequency as present in the stimulation signal. However, this signal also presents the higher-order harmonics. In some cases, the amplitude of the harmonic signal especially the third, and in some cases the fifth, harmonic is found larger than the amplitude of the fundamental. This variation was observed for different frequencies and in different people.

## 2.11  EEG Recorder

The EEG recorder used in this work was the BrainNet BNT 36, made by LYNX. The BNT - 36 also has an RJ45-type output for computer communication via the Ethernet network. The signal conditioning is done through instrumentation amplifiers with differential input and common mode rejection of more than 90 dB. For each channel the apparatus can filter signals in the bands of 0; 1; 0; 5; 1; 2; 5; 10; 20 and 50 Hz by means of high pass filter and signals in the bands of 20; 35; 70; 100 Hz through a low pass filter. In addition to the option to use the centralized Notch, filter at the frequency of 60 Hz. The signals after filtration can be sampled at rates of 100; 200; 240; 300; 400; 600, samples per second, in this work we opted for a sample rate of 600 Hz.

In conjunction with BrainNet, we chose to use silver electrodes, more precisely silver chloride Ag/AgCl. The Ag/AgCl electrodes have approximately 8 mm in diameter and conductive wire of 1.5 m.

## 2.12  Exam Protocol

The protocol was followed in each examination being important to try to adjust the conditions in which the user must operate the system. Parameters such as stimulus frequency, acquisition rate, commands to users, minimum distance between the stimulus source and the user should be observed before the start of the exams and defined as a protocol. In this work, the following parameters were adopted:

- Sampling frequency: 600 Hz, in order to work with a fixed window of 600 points;
- Channels used to calculate multiple coherence: $O1; O2; Oz; P3; P4; Pz$;
- Window size 600 points;
- Minimum distance of the stimulus to the user of 30 cm and maximum of 60 cm;
- Stimulation time from 10 s to 60 s.

The expected results for the protocol presented to each individual is shown in Fig. 3. This figure was created from the number of rotations of the motor of the robotic arm clamp.

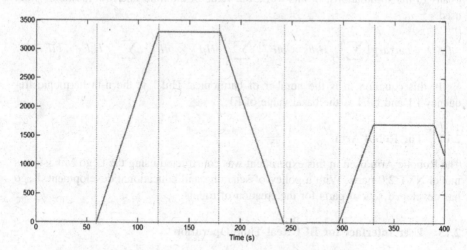

**Fig. 3.** Data from the encoder coupled to the robotic arm clamp motor.

The examination protocol had eight time intervals:

1. Interval, lasting 50 s, the individual should pay attention to no stimulus;
2. Interval, lasting 70 s, the individual should pay attention on the stimulation of 5 Hz. The engine would rotate clockwise;
3. Interval, lasting 50 s, the individual should pay attention to no stimulus;
4. Interval, with interval of 70 s, the individual should pay attention in the stimulation of 7 Hz. The engine would rotate counterclockwise;
5. Interval, with interval of 40 s, the individual should pay attention to no stimulus;
6. Interval, lasting 35 s, the individual should pay attention on the stimulation of 5 Hz. The engine would rotate clockwise;
7. Interval, lasting 50 s, the individual should pay attention to no stimulus;
8. Interval, with interval of 10 s, the individual should pay attention in the stimulation of 7 Hz. The engine would rotate counterclockwise;

## 2.13 Harmonic Sum Decision

After the data was processed through Multiple Coherence, the decisions were made using the Harmonic Sum Decision (HSD). This criterion uses not only the value of the fundamental component but also the harmonics from these frequencies mu05. The use of this type of decision maker stems from the fact that the evoked signal generates a response that differs between individuals. Because of this in some cases, the response obtained by the classifier was more pronounced in the second harmonic and not in the first one. Another reason is that the responses of different frequencies also vary the amplitude of coherence i.e. the response of a stimulus of 5 Hz has a greater amplitude than a signal of 13 Hz. Therefore, we chose to use the HSD that normalized the responses based on a baseline value, measured when the user was not paying attention to any of the stimulations. In this work, the value of the third and fifth harmonic ware used.

$$HSD = maxarg\left(\sum_{i=1}^{n} H_iF_1 \times blF_1^{-1}, \sum_{i=1}^{n} H_iF_2 \times blF_2^{-1}, \sum_{i=1}^{n} H_iF_3 \times blF_3^{-1}\right)$$

In this equation, n is the number of harmonics, HnF1 is the n-th harmonic frequency F1 and blF1 is the basal value of F1.

### 2.14  The Robot Arm

The Robotic Arm used in this experiment was constructed using the Lego Mindstorms model NXT 2:0 parts. With a policy of scientific and educational development, Lego has developed sets of parts for the creation of robots.

### 2.15  User Interface for BCI Real-Time Operation

In order to have an easy interface for the user to operate the system in real time, we developed in the MatLab GUI (Graphical User Interface) a program environment, with all the necessary commands for the acquisition of the data of the EEG apparatus without the need to use the program provided by the manufacturer. In this interface, the operator indicates the sampling frequency (100; 150; 200; 240; 300; 400; 600) in hertz. The number of points per window with the sampling frequency. The channel to be used as reference (the ground was adopted at the midpoint between Fp1 and Fp2), whether or not the data will be saved to file (.txt extension), the restart time of the MMSC, the set of channels to be used and the significance index ($\alpha$). As an output to the operator, the system indicates: if it is calibrated (if the basal values were calculated), the iteration number (corresponds to the total number of windows), the detected frequency value are sent to LabVIEW for moving the mechanical arm.

## 3   Results

Figure 4 shows the BCI develop in this work when in operation. The exams were performed on eight volunteers aged between 19 and 28 years, with normal vision or with corrections lenses and declared not to have any type of epilepsy.

To analyze the efficiency of the BCI, the correlation between the generated profile (protocol) and the response obtained by each volunteer. Table 1 presents the analysis results of these test data of the volunteers.

It can be seen that there is a better individual performance around the third and fourth exams. A conjecture about this fact is that there is a moment of training in the first exams that will culminate in the best answers (exams 3 or 4). After the best results, there was a drop in performance, which may be related to some type of visual or mental

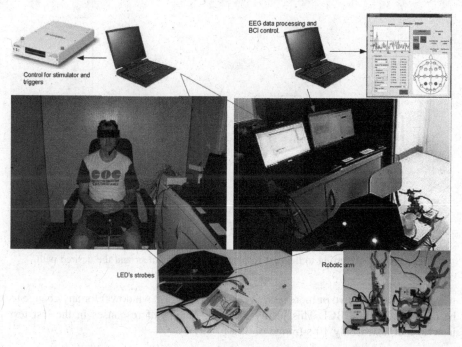

**Fig. 4.** Example of the complete operation of the proposed BCI.

**Table 1.** Correlation between the generated profile (protocol) and the response obtained by each volunteer. V/E stands for Volunteer/Exam

| V/E | 1 | 2 | 3 | 4 | 5 | 6 | 7 |
|---|---|---|---|---|---|---|---|
| 1 | 0.49 | 0.54 | 0.69 | 0.85 | 0.82 | 0.78 | 0.76 |
| 2 | 0.37 | 0.49 | 0.51 | 0.87 | 0.75 | 0.71 | 0.61 |
| 3 | 0.55 | 0.56 | 0.84 | 0.75 | 0.68 | 0.37 | – |
| 4 | 0.21 | 0.56 | 0.81 | 0.71 | – | – | – |
| 5 | 0.32 | 0.48 | 0.86 | 0.42 | 0.34 | – | – |
| 6 | 0.19 | 0.31 | 0.33 | 0.72 | 0.30 | – | – |
| 7 | 0.50 | 0.77 | 0.65 | 0.50 | – | – | – |
| 8 | 0.29 | 0.57 | 0.80 | 0.89 | 0.61 | – | – |

fatigue. The best test results are shown in Fig. 5. It presents the curves with the highest correlation value associated with each of the volunteers. It is possible to notice that, none of the exams could follow exactly the whole profile (protocol). However it is possible to notice that the trend presented in the profile was followed. Thus, it is noted that the motor movement controls have been given in the correct direction.

By analyzing the region where the system begins responding to the first concentration command (beginning of the second minute) (Fig. 5), it is possible to observe that this BCI takes about 10 to 15 s to initiate the response. Another fact worth mentioning is that every iteration of the BCI processing routine takes about 2 s. In this

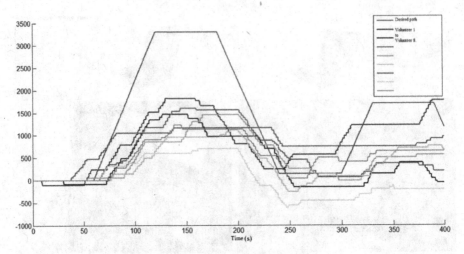

**Fig. 5.** Curves referring to the best response of each volunteer and the desired path.

way, it takes at least two or more seconds (1200 points or 2 windows) for any change to be perceived by the BCI. This justify the poor quality of responses in the last test interval that lasted only 10 s (range of 389 to 399 s).

## 4  Discussion and Conclusion

In this work, the study and development of a Brain Computer Interface was performed, using evoked potentials through intermittent photo stimulation. This photo stimulation was done using two light sources that blinked independently and with different frequencies. As the power of each evoked signal is related to its frequency [11], frequencies classified as low by [24] (frequencies in the range of 4 to 12 Hz) have been chosen. Based on these studies, it was chosen to work with the frequencies of 5 and 7 Hz that are within the range of 5 to 16.8 Hz pointed to as optimal by [25]. The signals were collected according to the international system 10–20 and the signals from the points P3 and 4 were used for the calculations. The combination of these channels produced a response higher than the responses of channels O1 and O2. The processing of the collected signals was done using the Multiple Magnitude Squared Coherence (MSC) in the analysis, where two channels ware used. A fact that was observed during data processing, was the presence of harmonics from the stimulation frequencies, these harmonics in some cases had a greater amplitude than the fundamental signal. This fact showed that the analysis of these signals could improve the result of the proposed Brain Computer Interface. In the work of [11], the authors demonstrate a formulation that takes into account the fundamental frequency signal amplitude and its harmonics (HSD - Harmonic Sum Decision). In the mentioned work, the authors used the data of the first three harmonics. The method presented in this work, was used as a decision maker, and used only the second harmonic (in this case, the frequencies of 10 and 14 Hz). In order to observe the efficiency of the BCI a group of eight volunteers were asked to operate the system following the light commands of the proposed protocol. To quantify and compare the responses, the correlation between the desired response and the data obtained in the exams was used. Considering that, the correlation is an index of similarity between two signals and that has a response between $-1$ and 1 where the worst case is when its value is null. The proposed BCI presented responses with correlations close to 0.9. With intra-averages individuals up to 0.7 and standard deviations that reached 0.26. By observing such data (Table 1), it is possible to note the training factor, which is indicated by the gradual increase of the correlations. It is also possible to observe the factor fatigue and/or accommodation. As an example, the first line of the Table 1: (0.49; 0.54; 0.69; 0.85; 0.82; 0.78; 0.76) shows the training when the system response goes from 0.49 to 0.85, which represents an improvement of 43%. This improvement is followed by a decrease of approximately 10%, which may be related to the fatigue or accommodation related to the stresses during the execution of the tests. A more detailed analysis of the data showed that the processing time of BCI is about 10 s. This implies that, on average, the BCI takes 10 s to perceive the presence or disappearance of an evoked signal, to classify it and ultimately to trigger or stop a response. This processing time justifies the low quality of the responses regarding the last 10 s of the protocol, in which the system did not have time to respond. It is worth reminding that the purpose of the protocol was to test the response of the system in several possible ways, including at the most critical points of operation.

# References

1. Wolpaw, J.R., Birbaumer, N., McFarland, D.J., Pfurtscheller, G., Vaughan, T.M.: Brain-computer interfaces for communication and control. Clin. Neurophysiol. 113(6), 767–791 (2002)
2. Dugdale, D., Hoch, D., Zieve, D.: Amyotrophic Lateral Sclerosis. A.D.A.M. Medical Encyclopedia (2010)
3. Pfurtscheller, G., da Silva, F.L.: Event-related EEG/MEG synchronization and desynchronization: basic principles. Clin. Neurophysiol. 110(18), 42–57 (1999)
4. Allison, B., Pineda, J.: ERPs evoked by different matrix sizes: implications for a brain-computer interface (BCI) system. IEEE Trans. Neural Syst. Rehabil. Eng. 11(110), 3 (2003)
5. Kubler, A., Kotchoubey, B., Kaiser, J., Wolpaw, J., Birbaumer, N.: Brain-computer communication: unlocking the locked. Psychol. Bull. 127(358), 75 (2001)
6. Olson, B., Si, J., Hu, J., He, J.: Closed loop cortical control of direction using support vector machines. IEEE Trans. Neural Syst. Rehabil. Eng. 13, 72–80 (2005)
7. Gao, X., Xu, D., Cheng, M., Gao, S.: A BCI-based environmental controller for the motion-disabled. IEEE J. NSRE 11(2), 137–140 (2003)
8. Wu, Z., Lai, Y., Xia, Y., Wu, D., Yao, D.: Stimulator selection in SSVEP based BCI. Med. Eng. Phys. 30(8), 1079–1088 (2008). Special Issue (part): Bioengineering in Taiwan
9. de Sa, A.M., Felix, L., Infantosi, A.: A matrix-based algorithm for estimating multiple coherence of a periodic signal and its application to the multichannel EEG during sensory stimulation. IEEE Trans. Biomed. Eng. 51(7), 1140–1146 (2004)

10. de Sá, A.M.F.L.M., Felix, L.B.: Improving the detection of evoked responses to periodic stimulation by using multiple coherence, application to EEG during photic stimulation. Med. Eng. Phys. **24**(4), 245–252 (2002)

11. Muller-Putz, G., Pfurtscheller, G.: Control of an electrical prosthesis with an SSVEP-based BCI. IEEE Trans. Biomed. Eng. **55**(1), 361–364 (2008)

12. McFarland, D.J., Wolpaw, J.R.: Sensorimotor rhythm-based brain computer interface (BCI): model order selection for autoregressive spectral analysis. J. Neural Eng. **5**(2), 155 (2008)

13. Wolpaw, J.R., McFarland, D.J., Vaughan, T.M., Schalk, G.: The wadsworth center brain-computer interface (BCI) research and development program. IEEE J. NSRE. **11**(2), 1–4 (2003)

14. Kim, H.K., Park, S., Srinivasan, M.A.: Developments in brain-machine interfaces from the perspective of robotics. Hum. Mov. Sci. **28**(2), 191–203 (2009)

15. Curran, E.A., Stokes, M.J.: Learning to control brain activity: a review of the production and control of eeg components for driving brain-computer interface (BCI) systems. Brain Cogn. **51**(3), 326–336 (2003)

16. Neuper, C., Muller, G.R., Kubler, A., Birbaumer, N., Pfurtscheller, G.: Clinical application of an EEG-based brain-computer interface: a case study in a patient with severe motor impairment. Clin. Neurophysiol. **114**(3), 399–409 (2003)

17. Patil, P.G., Turner, D.A.: The development of brain-machine interface neuroprosthetic devices. Neurother. Device Ther. **5**(1), 137–146 (2008)

18. Niedermeyer, E., da Silva, F.: Electroencephalography, Basic Principles, Clinical Application and Related Fields. Lippincott Williams & Wilkins (2005)

19. Vialatte, F., Maurice, M., Dauwels, J., Cichocki, A.: Steady-state visually evoked potentials: focus on essential paradigms and future perspectives. Prog. Neurobiol. **90**(4), 418–438 (2010)

20. Felix, L.: Detecção objetiva de respostas multivariável aplicado ao eletroencefalograma e a potenciais evocados, Master's teses, Programa de Pós-Graduação em Engenharia Elétrica, Universidade Federal de Minas Gerais, Belo Horizonte, Brasil (2004)

21. Wu, Z., Lai, Y., Xia, Y., Wu, D., Yao, D.: Stimulator selection in SSVEP based BCI. Med. Eng. Phys. **30**(8), 1079–1088 (2008). Special Issue (part): Bioengineering in Taiwan

22. Herrmann, C.: Human EEG responses to 1–100 HZ flicker: resonance phenomena in visual cortex and their potential correlation to cognitive phenomena. Exp. Brain Res. **137**(3), 346–353 (2001)

23. Zhang, Y., Jin, J., Qing, X., Wang, B., Wang, X.: LASSO based stimulus frequency recognition model for SSVEP BCIs. Biomed. Sig. Process. Control **7**(2), 104–111 (2011)

24. Zhu, D., Bieger, J., Molina, G.G., Aarts, R.M.: A survey of stimulation methods used in SSVEP-based BCIs. Comput. Intell. Neurosci. 12 (2010)

25. Bakardjian, H., Tanaka, T., Cichochi, A.: Optimization of SSVEP brain response with application to eight-command brain-computer interface. Neurosci. Lett. **469**, 34–38 (2010)

# An Adaptive User Interface Based on Psychological Test and Task-Relevance

Jaime A. Riascos[✉] ⓘ, Luciana P. Nedel ⓘ, and Dante C. Barone

Institute of Informatics, Federal University Rio Grande do Sul,
Porto Alegre, RS 91501-970, Brazil
jarsalas@inf.ufrgs.br

**Abstract.** The current advances into Human-Computer Interaction (HCI) are accompanied by the quick growth of Machine Learning (ML) and Artificial Intelligence (AI). It opens countless possibilities to enhance the interaction and communication between the machines and humans. Nevertheless, the human does not process the information at same way of computers (step-by-step); thus, the interfaces must be adequate to human's capabilities and limitations. The Graphical User Interfaces (GUIs) are the most used way to provide information to the user because the visual information is a straightforward and natural method to interact with the human. A new step forward is integrated some intelligent behavior for adapting the GUIs to the context, user's necessities and personalization; hence, the Adaptive User Interfaces (AUIs) are a new proposal which integrates intelligent and adaptive capabilities to achieve an enhanced human-computer interface. In this paper, an AUI is proposed based on psychological tests, which are used to assess the cognitive load that interface produces to the user so to know how much information must be shown and thus the user does not undergo overloading. These cognitive data is joined with a relevance weight, which represents what information is necessary to show in a given situation (Task-Relevance), in order to give to the user the most important data with the less cognitive load. The paper's aim is to show how psychological tests can be used as input information for an adaptive interface; in that sense, each them are explained and a possible user case is shown.

**Keywords:** Adaptive User Interface · Psychological tests · Task-relevance

## 1 Introduction

Since Card et al. [1] introduced the psychological knowledge into Human-Computer Interaction (HCI), several factors about the human behavior and processing information were taken into account to design and evaluate any interaction with machines [2].

Into HCI, Graphical User Interfaces (GUIs) has been the main way to handle communication between the users and machines. Initially, the guidelines for user interfaces design came from the human psychology: how our perception, memory and cognition work [3]; therefore, the human capabilities and limitations play an important role in HCI. In GUI, it is necessary to reach two main goals: giving clearly some information (data visualization) and responding to a given instruction (interaction).

© Springer International Publishing AG 2017
D.A.C. Barone et al. (Eds.): LAWCN 2017, CCIS 720, pp. 143–155, 2017.
https://doi.org/10.1007/978-3-319-71011-2_12

It is clear that humans have a limited capacity to process information. The Model Human Processor [1] explains how the information flows through our mind and how it is processed; thus, in order to make the interfaces more usable, their design must be based on these limitations (human factors) [4, 5].

The traditional GUIs are limited on its functionality. They were simply designed to statically show information and do not take into account the personalization by the user and the context adaptation [6]. A new approach to face these issues is the Adaptive User Interface (as well-known Intelligent User Interface – IUI) [7]. This method uses artificial intelligence techniques to improve the interaction through the adaptation of the display based on the abilities, limitations, personalization, and context of the user.

This work proposes an Adaptive User Interface (AUI) where the adaptation rule is determined by both the cognitive load of the screen and the relevant information that must be showed (Task-Relevance). In one side, following approaches as [8, 9], where the menu's performance and complexity of an interface were measured respectively. The cognitive load of each item or object on the proposed interface is represented by a weight. This measure can be obtained from psychological tests such as Electroencephalography (EEG), Heart Rate, Eye Tracking and Pupillometry, Electrodermal Activity (EDA) [10].

On the other side, a relevance factor can be obtained from both the prior knowledge about the operation and actions performed by the user on the system. Based on Fuzzy Logic, it is possible to handle different levels of relevance so that they can be joined to the above weight and thus showing the most relevant information with the less cognitive load. These data are the input for a Bayesian Network (BN) which is the proposed method for the making decision.

The structure of this paper is as follows: Sect. 2 presents the related works. Section 3 explains the AUI based on human processing information and its functionality. Section 4 introduces the psychological test which can be used to measure the cognitive load. Section 5 shows a possible user case where the proposed interface could be applied. Finally, Sect. 6 concludes the paper and presents futures directions.

## 2   Related Works

In a few decades ago, the fast grow of Machine Learning (ML) has made that several authors play attention on AUIs. It is demonstrated by Cortes et al. [6] how AUIs have improved the communication and interaction with machines as well as many disciplines such as cognitive sciences, ergonomics, computer sciences have influenced in the AUI's development. Likewise, many authors have discussed the necessity of a new generation of interfaces that can have some kind of intelligence and thus enhancing the human-machine communication [11–13].

In that sense, several researchers have worked on either items or menus adaptation, mainly their distribution and removing on the interface. In early works, Shneiderman and Mitchell [14] show a dynamical system to order the interface's menus based on the frequency of use. Similarly, Cockburn et al. [8] developed a model that predicts the performance of many different menu designs. This model is based on Fitts' law and Hick-Hyman law. These laws are representatives on human factors and human processing information [2]. Likewise, Miniukovich and De Angeli [9] quantify the visual complexity of an interface using three dimensions: Information amount, Information Organization, and Information discriminability. They finally proposed six associated automatic metrics to evaluate their approach. Although the above authors argued methods for assessing performance on interfaces, there is no a discussion about the user's cognitive load; thus, in this paper is proposed to individually measure each item on the interface to quantify its cognitive load and taking it into account to decide when, how and which items must be shown.

Due to AUIs are an user-centered design (UCD) approach [15], the creation of the user model plays an important role. For instance, Kiencke et al. [16] use the Model Human Processor (MHP) to build a structural model of the driver behavior. In this paper, the MHP is used as a start point to understand the human capabilities and limitations and hence the development of the interface is performed using these factors. Another integration of human abilities for designing AUIs was carried out by Zudilova-Seinstra [17]. In that implementation, the user model was performed using the Wagner's Ergonomic Model, which classifies the human factors to characterize the user. However, these approaches make their deductions based on the human behavior interacting with the interface. Thus, psychological measures as cognitive load or stress levels were not used. In this paper are proposed several techniques to collecting these measures and thus they can be used as input data for performing an AUI.

Gullà et al. [15] performed a proposed of AUIs to manage the modeling information of the user in a daily domestic situation. They employed three sources of information to perform the AUIs: User, Environment, and Interaction models. Moreover, they proposed the use of Bayesian Network (BN) as the adaptation mechanism. Likewise, in another approach, Gullà et al. [18] demonstrated as BN is a well-established method for decision-making in AUIs.

## 3 Adaptive User Interface

The common AUIs have taken into account at least the following three key factors to [6, 11]:

- User Modelling
- Adaptive Engine
- Interaction Repository

### 3.1 The User Modelling

It refers to physical and mental aspects, behaviors and preferences of the user; a well-known model is the proposed by Rasmussen [19], where the human behavior is divided in skill, rule, and knowledge. Although the Rasmussen' model can describe very well the user needs, the interfaces are also designed based on the human abilities and limitations; therefore, understanding how human processes the information is an essential key to develop useful interfaces. The model human-processor (Fig. 1.) introduced by Card et al. [1] synthesizes how the information flows through of three processors (Perceptual, Cognitive and Motor) before the human make an action (motor response).

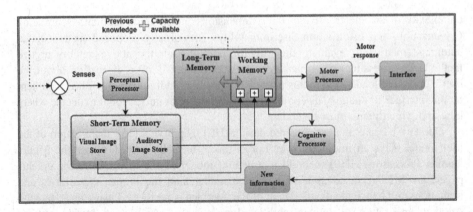

**Fig. 1.** Model human processor. Adapted from [1].

First, the new information comes into Perceptual Processor through sensory input to be quickly held in the Short-Term-Memory [5]. The human's capacity for remembering was established by Miller [20] as seven ± two chunks of information. In that sense, the information visualization on the interface can be seen as resource management where the number objects were shown must be optimized in function of the human limitations.

The input information is carried to Cognitive Processor where a decision is made; Working Memory processes and matching the new information with the stored in the Long-Term Memory and thus to choose the action to do (make decisions). To avoid the ambiguity around the concepts of short-term memory and working memory, it was assumed that short-term memory is part of the sensory store and working memory is in charge of manipulating information and making operations (a deep discussion can be found in [21]).

Throughout of information processing can be identified the Reaction Time (RT) and Movement Time (MT) [22]. The first one depends on the complexity of information to be processed; it can be measured since the stimuli begin until the motor response. In addition, RT varies according to channel sensory. MT depends largely on psychical characteristics of the user. Then, the interface must be thought to try to improve the performance (right choice) and reducing RT.

Normally, the user-interface interaction is a rich source of information about the user behavior. In order to make an interface more natural and personalizable, the proposed interface creates a user model based on that data and the user preferences. The model human processor is used as a constraint in the design process, where measures as RT gives a continuous feedback about the interface's performance and thus the model modifies itself in each use.

## 3.2 Adaptive Engine

It is in charge to make the settings, decisions, and predictions. The core of interface's adaptive behavior is the knowledge base; it comes from the data and interaction of user [17]. Therefore, in this paper, the knowledge base is mainly made of the interface's load, the importance level of information that is being shown (tasks-relevance) and the interaction user-interface (personalization). For the first item, the interface's load ($\lambda$) is defined as:

$$\lambda = \sum_{i=0}^{n} k_i \qquad (1)$$

where $k$ is the cognitive load for each object on the interface. It is supposed that the value of $n$ is not greater than the human capacity (constraint). How to obtain the cognitive values is explained in the next section (psychological tests).

The importance level is though as a weight ($W$) which is obtained using Fuzzy Logic (FL) [23]. In a given process with several inputs, some variables at an instant $t$ are more important to show than others, e.g. an emergency state; hence, the user, based on its experience, establishes the rules for the FL. Thus, in each instant $t$ there will be a number of objects $n$ that were chosen based on their importance. In this point, it is necessary a trade-off between the importance of the shown data and human capabilities. Because it is so important to show the relevant information for the user, but if this implies an overload to user, the interface becomes useless. Thereby, for each $k$ is associated a $W$ so that in an instant $t$ the corresponding objects to be shown. Then, a load factor ($\omega$), which associates the cognitive load with the weight, can be write as:

$$\omega = \sum_{i=0}^{n} k_i W \qquad (2)$$

The personalization aspect plays an important role into the AUI, it is tough to handle the controllability of the interface, where the user can decide an initial customization and setting up different both objects or disposition. Therefore, the personalization is an independent factor and it does not influence on the adaptive loop.

Into the Adaptation Engine, the Bayesian Network (BN) [24] is the method to manage the AUI, each node is a state or variable into the network and the links are the direct relationship among the nodes. Moreover, the default probabilities of the BN are reinforced by the Eq. 2.

### 3.3   Interaction Repository

Finally, the interaction repository stores the interface's objects and parameters which can be adjusted either by the user (manually) or computer system (automatically). The creation of this repository starts with a static interface (GUI), where the designer creates any objects, charts, graphs, bottoms, etc. Thus, each of them is presented to the user to be evaluated. With this initial evaluation can start the Adaptive Interface loop; the Fig. 2 summarizes the discussion about the Adaptive User Interface.

## 4   Psychological Tests

While the human is performing a cognitive task (learning, solving problems, so on), he experiments a workload to do so. Initially, the quantification was assumed based on the empirical examination of the relationship between problem-solving and learning [10].

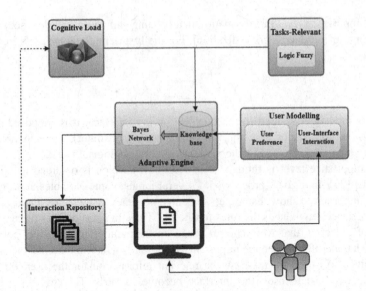

**Fig. 2.** Adaptive interface loop.

Now, the technological advances allow doing measures more directly related to mental states (stress, anxiety, and fatigue). In this section, several methods to obtain the $k$ value (cognitive load for interface's objects) are presented. It is possible to use one or more of them to obtain the cognitive load to be used in the adaptive engine.

## 4.1   EEG

The Electroencephalography (EEG) is a non-invasive method to obtain the electrical activity of the neurons. The data recorded is processed to decode the user's mental state. This technique has widely been used for cognitive measurements because it is possible to identify various types of load instead only the overall load [10]. The data analysis represents the most important step in the EEG method. Indeed, the great amount of information (spatial, spectral and temporal) is a disadvantage of this method, which can lead to confusing the real user's state. Likewise, this technique is often uncomfortable for the user, implying an amount of stress by the entrance.

Commonly, the prefrontal cortex is associated to Working Memory [25]. Klimesch [26] highlights how Alpha (7.5–12.5 Hz) and Beta (4–7.5 Hz) wave rhythms have been used for analyzing the cognitive and memory performance. Likewise, Kumar and Kumar [27] explain a step-step for extracting the cognitive load using EEG through of a power spectrum analysis. The Fig. 3 shows the experimental setup that evaluation. Moreover, Anderson et al. [25] make a deep explanation of how to use EEG for studying cognitive load in a visualization experiment.

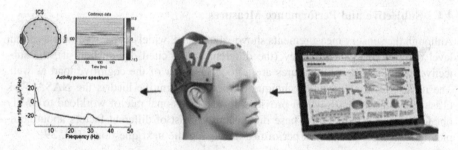

**Fig. 3.** Example of an interface using EEG. Took from [27].

## 4.2  Eye-Tracking

The Eye-tracking consists on capturing the eye gaze for calculating the point where the user is looking at (point of regard).

There are two techniques to perform Eye-tracking: video-based and EOG (Electrooculography) based [28]. The video base uses tracking cameras to capture the pupil movement and the scene field. The typical sampling rate is 120 Hz and it is possible to reach accuracies of 0.5°. Eye-tracking based on EOG employs two pairs of skin electrodes to get the electrical signal produced by the eye's movements, likewise as in EEG. A deep discussion of these techniques can be found in [28, 29].

For understanding the cognition and mental behavior of the user, there are three main types of information about the eye that can be used: Saccades, Fixations, and Blinks. The saccades are the rapid movement of the eyes to change a region of interest to other in the visual scene. Fixations are such stationary states where the user is looking at and holding its attention and finally the blinking is a rapid semi-conscious open-closing eyelid to keep the eye lubricated. Several authors had successfully used these types of eye information to study and monitoring the perceptual and cognitive load of the user during tasks [30, 31] and EEG combined with some eye-tracking technique [32, 33].

Another new approach has been used to measure the cognitive load based on pupil information. The Pupillometry employs the pupil size to estimate the induced load for a task to the user [34, 35]. The pupil sensitively responses (dilates) to both visual and non-visual stimulus and it corresponds to workload levels.

## 4.3  Electrodermal Activity (EDA)

In this approach, the electrical activity of the skin in response to sweat secretion is used to indicate stress, emotion or workload. It had been demonstrated how EDA indexes change depending on the mental states [36]. The type of skin responses can be Skin Resistance Level (SRL), Galvanic Skin Response (GSR), Skin Conductance Response (SCR), among others. Setz et al. [37] demonstrate, in an experimental setting, how to estimate the cognitive load and stress level using a wearable EDA device to measure the SCR. Nourbakhsh et al. instead SCR, use GSR to estimate the cognitive load in arithmetical and reading tasks [38].

## 4.4    Subjective and Performance Measures

Although the sensory measurements shown above have widely been used with a certain level of feasibility and accuracy (the different works cited demonstrate it), the subjective and performance measures are a good indicator of the cognitive load as well. The most typical subjective technique to measure cognitive load is the NASA Task (Task Load Index) [39], which provides a six-dimensional rate of workload to finally obtain an overall workload. These dimensions consist of different factors about completing a task. NASA task is performed answering the next questions: [10, 39]:

- How much mental and perceptual activity was required? (Mental demands).
- How much was physical activity required? (Physical demands).
- How much time pressure occurred? (Temporal demands).
- How successful do you think you were in accomplishing the goals of the task set by the experimenter? (Performance).
- How hard did you have to work –mentally and physically- to accomplish your level of performance? (Effort).
- How insecure, discouraged, irritated, stressed versus secure, content and relaxed did you feel during the tasks? (Frustration level).

Even though the NASA-TLX yields a useful summary about the complexity level of the tasks for the users, it cannot give ongoing information about the current state of the user performing the task; Also, the limitation of NASA is that only can be made once per task. Even so, Hart [40] points out that the NASA-TLX is very useful to evaluate the interface design.

Furthermore, the performance metrics are a common metrics to assess a task. The Reaction Time (RT) and Error Rate (ER) related to time to take to the user to make an action and the right choice expected respectively [22]. The most complex work would be to establish what it is a correct choice (error definition) and what the start and end time points will be used to measure the RT. Thus, a higher cognitive load is often presented with a longer RT and high ER [10].

Other measurements that were not explained (and not less important) on this section as Heart Rate, fNIRS, fMRI, respiration, EMG, among others, can be used to estimate the cognitive load as well. Hence, Fairclough et al. [41], Anderson et al. [25], Jiao et al. [42] demonstrate the feasible use of several psychological signals (as the explained above) to measure the cognitive load and mental states.

Nevertheless, the correct method to obtain the cognitive load must be accompanied by an adequate task design so that a psychological state can be clearly detected. In order to reach it, a baseline state must be established [43] to identify a real change on the mental state. Among the different types of baseline, the dual-task methodology is the most common [10], where a primary task is defined as a reference state to be used for comparison when a secondary task is added. Thus, a validated response to the task will be interpreted in any significant deviation from the resting state [43]. In the work of Ziegler et al. [44] can be found a study using secondary tasks and psychological tests.

## 5  User Case

The GUIs are an essential part of monitoring and inspection systems. However, the interfaces normally yield a lot of information that overload the user [5]. The supervision of an ongoing operation is a typical case. In that sense, a teleoperation robotic system was simulated to start with the implementation In this first stage, the static interface uses simulated numerical data of a robotic operation (inspection through a pipeline), where three variables were considered (velocity, rotation and deep). Each object in the interface was defined based on an end-user approach. Thereby, an initial model of the interface is given for the creation of the interaction repository and thus it starts its evaluation using any method explained above. The Figs. 4 and 5 show the two interfaces used on a screen each. On the Screen A, the global information about the operation is presented, and on the Screen B the local information (physical robot state).

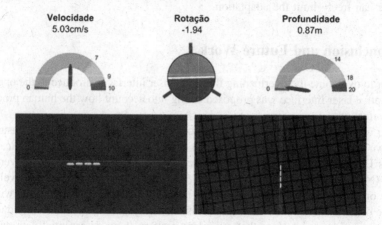

**Fig. 4.** Screen A – user interface.

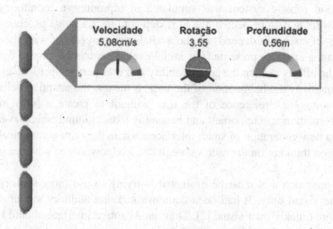

**Fig. 5.** Screen B – user interface.

Then, the user can set the default disposition that the interface must start, it is based on the relevance of the information that is shown in each the object (Task-Relevance), e.g. in an emergency state, the priority is looking the velocity instead of the deep.

Following the dual-task approach, when the user sets up the default interface, it is explicitly establishing a baseline interface to start with the cognitive load recordings of the objects. After choosing what type of psychological test will be used (past section), the simulation of operation starts and the cognitive load of baseline state can be collected. From here, the cognitive load of every Task-Relevance and new objects shown are measured in comparison with the baseline established. Thus, so that the adaptation cycle can start, it is necessary that at least once the simulation ran completely to obtain the enough cognitive load data.

Finally, the adaptation cycle can start with the cognitive and the user interaction data collected. Here, the personalization aspect plays an important role in the adaptation, because the user can have preferences with some data or object; for that reason, the user can freely limit the adaptation.

## 6 Conclusion and Future Work

In order to overcome the overloading that static user interfaces produce to the operator, an Adaptive User Interface was proposed taking into account how the human processes the information.

The current proposal has two main faces: the first one is trying to measure the cognitive load that the interface has. In that sense, several types of psychological tests such as EEG, Eye-tracking, and EDA were presented and described; moreover, subjective (NASA-TLX) and performance (RT, ER) methods were explained as well. The second one is the Task-Relevance, which is set up beforehand by the user with the purpose of showing the data only when it is necessary. Thus, the aim is to reduce the overload creating an interface that can adapt itself to show an essential data with the less cognitive load.

Likewise, this paper shows the first stage of this proposal with a suitable user case. A teleoperation robotic system was simulated to reproduce a monitoring task. In accordance with the path presented, the next step is to choose what psychological tests will be used; Of course, it depends on the available resources, but with the subjective and performance methods presented, an initial evaluation can be carried out.

Despite the efforts to keep the controllability and user personalization, an important issue that remains is the habituation of the user; it means, the adaptive behavior must take into account the preference of the user without to create a habit making the relevance information less important and becoming a background task. Nevertheless, it is necessary a new generation of smart interfaces, where they can automatically change their disposition thinking on the user's state. It can lead towards an advance in affective computing.

Another approach that must be evaluated is trying to use more sensory channels instead of the visual only. It had been demonstrated that auditory stimulus could be processed more quickly than visual [2]. Thus, an Adaptive Interface should know what type of stimuli to use depending on the situation.

**Acknowledgement.** Jaime A. Riascos is supported by a master degree studentship from FAURGS/Petrobras, research project 8147-7. The first author would like to thanks to Prof. Dr. Lewis Chuang for hosting him in his Lab and provide him a theoretical discussion about the subject of this paper. That internship was done by funding provided by the SFB TRR-161 (Work package C03).

# References

1. Card, S.K., Newell, A., Moran, T.P.: The Psychology of Human-Computer Interaction. L. Erlbaum Associates Inc., Hillsdale (1983)
2. Seow, S.C.: Information theoretic models of HCI: a comparison of the Hick-Hyman law and Fitts' law. Hum.-Comput. Interact. **20**(3), 315–352 (2005)
3. Johnson, J.: Designing with the Mind in Mind: Simple Guide to Understanding User Interface Design Rules. Morgan Kaufmann, Boston (2010)
4. Hall, S.K., Cockerham, K.J., Rhodes, D.J.: Applying human factors in graphical operator interfaces. In: Annual Pulp and Paper Industry Technical Conference, Portland, USA (2001)
5. Helander, M.: A Guide to Human Factors and Ergonomics. CRC Press Taylor & Francis Group, London (2006)
6. Cortes, V.A., Zarate, V.H., Ramirez Uresti, J.A., Zayas, B.E.: Current challenges and applications for adaptive user interfaces. In: Human-Computer Interaction, pp. 49–68. InTech (2009)
7. Rothrock, L., Koubek, R., Fuchs, F., Haas, M., Salvendy, G.: Review and reappraisal of adaptive interfaces: toward biologically inspired paradigms. Theor. Issues Ergon. Sci. **3**(1), 47–84 (2002)
8. Cockburn, A., Gutwin, C., Greenberg, S.: A predictive model of menu performance. In: Proceedings of the SIGCHI Conference on Human Factors in Computing Systems, San Jose, California, USA (2007)
9. Miniukovich, A., De Angeli, A.: Quantification of interface visual complexity. In: Proceedings of the 2014 International Working Conference on Advanced Visual Interfaces - AVI 2014, Como, Italy (2014)
10. Sweller, J., Ayres, P., Kalyuga, S.: Measuring cognitive load. In: Sweller, J., Ayres, P., Kalyuga, S. (eds.) Cognitive Load Theory, pp. 71–85. Springer, New York (2011). https://doi.org/10.1007/978-1-4419-8126-4_6
11. Mezhoundi, N., Khaddam, I., Vanderdonckt, J.: Toward usable intelligent user interface. In: Kurosu, M. (ed.) HCI 2015. Springer, Cham (2015). https://doi.org/10.1007/978-3-319-20916-6_43
12. Zaki, M., Forbrig, P.: Towards the generation of assistive user interfaces for smart meeting rooms based on activity patterns. In: Paternò, F., de Ruyter, B., Markopoulos, P., Santoro, C., van Loenen, E., Luyten, K. (eds.) AmI 2012. LNCS, vol. 7683, pp. 288–295. Springer, Heidelberg (2012). https://doi.org/10.1007/978-3-642-34898-3_19
13. Cerny, T., Chalupa, V., Donahoo, M.: Towards smart user interface design. In: International Conference on Information Science and Applications, Suwon (2012)
14. Shneiderman, B., Mitchell, J.: Dynamic versus static menus: an exploratory comparison. SIGCHI Bull. **20**(4), 33–37 (1989)
15. Gullà, F., Cavalieri, L., Ceccacci, S., Germani, M., Bevilacqua, R.: Method to design adaptable and adaptive user interfaces: a method to manage the information. In: Stephanidis, C. (ed.) HCI 2015. CCIS, vol. 528, pp. 19–24. Springer, Cham (2015). https://doi.org/10.1007/978-3-319-21380-4_4

16. Kiencke, U., Majjad, R., Kramer, S.: Modeling and performance analysis of a hybrid driver model. Control Eng. Pract. **7**, 985–991 (1999)
17. Zudilova-Seinstra, E.: On the role of individual human abilities in the design of adaptive user interfaces for scientific problem-solving environments. Knowl. Inf. Syst. **13**, 243–270 (2007)
18. Gullà, F., Cavalieri, L., Ceccacci, S., Germani, M.: A BBN-based method to manage adaptive behavior of a smart user interface. Procedia CIRP **50**, 535–540 (2016)
19. Rasmussen, J.: Skills, rules, and knowledge; signals, signs, and symbols, and other distinctions in human performance models. IEEE Trans. Syst. Man Cybern. **SMC-13**(3), 257–266 (1983)
20. Miller, G.A.: The magical number seven, plus or minus two: some limits on our capacity to process information. Psychol. Rev. **63**(2), 81–97 (1956)
21. Aben, B., Stapert, S., Blokland, A.: About the distinction between working memory and short-term memory. Front. Psychol. **3**, 301 (2012)
22. Dix, A., Finlay, J., Abowd, G.D., Beale, R.: Human-Computer Interaction. Pearson, Harlow (2004)
23. Zadeh, L.: Fuzzy logic. Computer **21**(4), 83–93 (1988)
24. Pearl, J.: Probabilistic Reasoning in Expert Systems: Networks of Plausible Inference. Morgan Kaufmann, San Francisco (1988)
25. Anderson, E.W., Potter, K.C., Matzen, L.E., Shepherd, J.F., Preston, G.A., Silva, C.T.: A user study of visualization effectiveness using EEG and cognitive load. Comput. Graph. Forum **30**(3), 791–800 (2011)
26. Klimesch, W.: EEG alpha and theta oscillations reflect cognitive and memory performance: a review and analysis. Brain Res. Rev. **29**, 169–195 (1999)
27. Naveen, K., Kumar, J.: Measurement of cognitive load in HCI systems using EEG power spectrum: an experimental study. Procedia Comput. Sci. **84**(1), 70–78 (2016)
28. Majaranta, P., Bulling, A.: Eye tracking and eye-based human-computer interaction. In: Fairclough, S.H., Gilleade, K. (eds.) Advances in Physiological Computing. HIS, pp. 39–65. Springer, London (2014). https://doi.org/10.1007/978-1-4471-6392-3_3
29. Flad, N., Fomina, T., Buelthoff, H.H., Chuang, L.L.: Unsupervised clustering of EOG as a viable substitute for optical eye tracking. In: Burch, M., Chuang, L., Fisher, B., Schmidt, A., Weiskopf, D. (eds.) ETVIS 2015. MV, pp. 151–167. Springer, Cham (2017). https://doi.org/10.1007/978-3-319-47024-5_9
30. Chen, S., Epps, J., Chen, F.: Automatic and continuous user task analysis via eye activity. In: Proceedings of the 2013 International Conference on Intelligent User Interfaces, Santa Monica (2013)
31. Chen, L., Pu, P.: Eye-tracking study of user behavior in recommender interfaces. In: De Bra, P., Kobsa, A., Chin, D. (eds.) UMAP 2010. LNCS, vol. 6075, pp. 375–380. Springer, Heidelberg (2010). https://doi.org/10.1007/978-3-642-13470-8_35
32. Plöchl, M., Ossandón, J., König, P.: Combining EEG and eye tracking: identification, characterization, and correction of eye movement artifacts in electroencephalographic data. Front. Hum. Neurosci. **6**, 278 (2012)
33. Flad, N., Bulthoff, H., Chuang, L.: Combined use of eye-tracking and EEG to understand visual information processing. In: International Summer School on Visual Computing (VCSS 2015), Stuttgart (2015)
34. Peysakhovich, V., Dehais, F., Causse, M.: Pupil diameter as a measure of cognitive load during auditory-visual interference in a simple piloting task. Procedia Manuf. **3**, 5199–5205 (2015)
35. Wong, H., Epps, J.: Pupillary transient responses to within-tasks cognitive load variation. Comput. Methods Programs Biomed. **137**, 47–63 (2016)

36. Critchley, H., Nagai, Y.: Electrodermal activity (EDA). In: Gellman, M.D., Rick Turner, J. (eds.) Encyclopedia of Behavioral Medicine, pp. 666–669. Springer, New York (2013)
37. Setz, C., Arnrich, B., Schumm, J., La Marca, R., Troster, G., Ehlert, U.: Discriminating stress from cognitive load using a wearable EDA device. IEEE Trans. Inf. Technol. Biomed. **14**(2), 410–417 (2010)
38. Nourbakhsh, N., Wang, Y., Chen, F., Calvo, R.A.: Using galvanic skin response for cognitive load measurement in arithmetic and reading tasks. In: Proceedings of the 24th Australian Computer-Human Interaction Conference, Melbourne (2012)
39. Hart, S.G., Staveland, L.E.: Development of NASA-TLX (Task Load Index): results of empirical and theoretical research. Adv. Psychol. **52**, 139–183 (1988)
40. Hart, G.S.: NASA-task load index (NASA-TLX); 20 years later. Proc. Hum. Factors Ergon. Soc. Annu. Meet. **50**(9), 904–908 (2006)
41. Fairclough, S.H., Moores, L.J., Ewing, K.C., Roberts, J.: Measuring task engagement as an input to physiological computing. In: 3rd International Conference on Affective Computing and Intelligent Interaction and Workshops, Amsterdam (2009)
42. Jiao, X., Bai, J., Chen, S., Li, Q.: Development of physical workload measuring system based on multi-physiological signals from human-machine interaction. In: Proceedings of IEEE International Conference on Virtual Environments, Human-Computer Interfaces, and Measurement Systems, VECIMS (2012)
43. Parchment, A., Wohleber, R.W., Reinerman-Jones, L.: Psychophysiological baseline methods and usage. In: Schmorrow, D.D.D., Fidopiastis, C.M.M. (eds.) AC 2016. LNCS, vol. 9743, pp. 361–371. Springer, Cham (2016). https://doi.org/10.1007/978-3-319-39955-3_34
44. Ziegler, M.D., Kraft, A., Krein, M., Lo, L.-C., Hatfield, B., Casebeer, W., Russell, B.: Sensing and assessing cognitive workload across multiple tasks. In: Schmorrow, D.D.D., Fidopiastis, C.M.M. (eds.) AC 2016. LNCS, vol. 9743, pp. 440–450. Springer, Cham (2016). https://doi.org/10.1007/978-3-319-39955-3_41

# Decision Trees

# Decision Tree to Analyses EEG Signal: A Case Study Using Spatial Activities

Narúsci dos Santos Bastos[1,2], Diana Francisca Adamatti[2(✉)],
and Cleo Zanella Billa[2]

[1] Universidade Federal de Pelotas, Pelotas, RS, Brazil
nds.bastos@inf.ufpel.edu.br
[2] Universidade Federal do Rio Grande, Rio Grande, RS, Brazil
dianaada@gmail.com, cleo.billa@gmail.com

**Abstract.** The Electroencephalogram (EEG) is based on records of brain electrical activity measured through the scalp, and it is commonly used for studies and neurological examinations. It has now been used in conjunction with BMI (Brain Machine Interface) Systems, which allow the communication of an individual and external equipment, such as a computer, through brain stimulation, without any muscle movement. These tools typically present data in the form of graphs or topographic maps for monitoring or analysis of brain activities. In this work, a methodology of EEG analysis is proposed through data mining, which tries to extract knowledge from a large database. This paper presents a case study using EEG signals from visually impaired and sighted individuals during the execution of an activity that stimulated spatial ability. In order to verify the hypothesis that sighted use the sense of sight, even with blindfold, and visually impaired use the sense of touch to identify spatial geometric objects. An experiment was made and through the data of the recorded brain signals, decision tree techniques were applied to understand the main areas involved in the brain activities during the execution of the recognition task. The results suggest that the hypothesis is true confirming that visuall and sighted subjects activate different brain areas.

**Keywords:** Decision tree · Data mining · Brain Machine Interface · Visual impairment · Electroencephalogram · EEG

## 1 Introduction

BMI (Brain Machine Interface) is an application that allows communication between an individual and an external device, such as computers, without any muscular movements created by the brain [13]. These devices make direct communication through brain impulses. The EEG (Electroencephalogram) is based on records of electrical brain activity that are measured on the surface of the scalp. These high temporal resolution systems, for example, are able to measure brain activity at every second miracle, generating a large amount of data. The human brain is a multifaceted structure, capable of storing large amounts of information, transforming them, learning and making complex decisions, providing us with the ability to discover and influence

D.A.C. Barone et al. (Eds.): LAWCN 2017, CCIS 720, pp. 159–169, 2017.
https://doi.org/10.1007/978-3-319-71011-2_13

the world around us [5]. In this sense, the branch of neuroscience is an interdisciplinary science that joins different areas of knowledge with the intention of interpreting the nervous system as a whole. In this way, different studies based on BMI systems are published and applied with varied purposes as to provide better quality of life to people with severe motor problems [15], rehabilitation of stroke victims [13], robotic systems movement, classification of teaching objects relating students level of attention [17], besides that, in conjunction with EEG equipment allows the analysis of the brain activity through images and topographic maps 2D and 3D and graphical analyzes.

In order to find a way of analyzing the brain activity that was different from the graphical analysis, we tried to analyze EEG data through decision tree. The decision tree presents as one of its most important characteristics the easier way to interpreter the rules inferred through them [1].

This work uses decision tree for the understanding of brain signals, for this a case study was applied with tasks that exploit the spatial ability from the identification of spatial objects of visually impaired and sighted people.

Visually impaired have their abilities of orientation, movement in space with compromised security and independence [8]. Therefore, it is very important to stimulate the capacities of the other sense organs, since from the other senses, the individual will adapt himself (herself) to the world [14]. Basing in this idea, a question arises: do sighted and visually impaired people activate the same brain areas during the recognition of a spatial object? Once we know that people who have the ability to see, have the occipital lobe responsible for the ability of vision, which the area is responsible for the spatial activity of a visually impaired?

Some works discuss this type of approach, however, they use data mining techniques for classification. However, this work does not involve classification of data, but rather a way to visualize and to interpret the data through decision trees. The literature researched does not present similar works.

This paper presents a methodology for the use of decision trees as a means to understand brain activity from a case study with the recognition of spatial objects. Section 2 presents part of the theoretical framework used for the development of the work. Section 3 shows the proposed methodology used. In Sect. 4, the case study and the generated models are presented and, finally, in Sect. 5 is the results of this research.

## 2 Theoretical Basis

### 2.1 Brain Areas and Their Functions

The brain is the main component of the nervous system. It is responsible for all mental operations such as concentration, thinking, learning and motor control. These capabilities are implemented through neurons, which can currently be explained by neuroscience.

Anatomists usually divide the brain into major regions, called lobes, whose boundaries are not always accurate, but transmit an initial idea of regional location. There are five lobes: four external and one internal, located in the lateral sulcus [11]. The four external lobes are: frontal, which is located in the forehead; parietal, which is located under the cranial bone with the same name; temporal, which is associated with the temple; and occipital, which is located in the occipital cranial bone. The fifth lobe, the insula lobe can only be seen when the lateral sulcus is opened [7, 11]. There are many other structures situated in the central nervous system (CNS), but in this work we investigate only the four visible lobes because the BCI system that we used do not have access to the insula lobe.

Each lobe has specialized functions: the occipital lobe is primarily concerned with the sense of vision, it is divided into multiple distinct visual areas, in which the biggest one is the primary visual cortex. The parietal lobe is partially dedicated to the sense of touch, it is responsible for body sensitivity functions and spatial recognition. The temporal lobe contains the primary auditory cortex, it processes audio data, specific aspects of vision, language understanding and some aspects of memory. Finally, the frontal lobe which is responsible for cognitive actions, memory and movement [7, 11]. Table 1 presents the brain areas and their main functions.

**Table 1.** Brain region, electrodes and proprietary functions

| Brain region | Electrode | Proprietary functions |
|---|---|---|
| Frontal lobe | FP1, FP2, FZ, F7, F3, F4, F8, FT9, FT10, FC5, FC1, FC2, FC6 | Executive functions (management of cognitive/emotional resources on a given task) |
| Temporal lobe | T7, TP9, T8, TP10 | Perception of biological motion |
| Parietal lobe | P7, P3, PZ, P4, P8, C3, CZ, C4, CP1, CP2, CP6, CP5 | Somatosensory perception, spatial representations and tactile perceptions |
| Occipital lobe | O1, OZ, O2 | View images (including during a dialogue) |

## 2.2 Decision Trees

The decision tree is a supervisioned classification technique based on the division of a complex problem into several subproblems, repeating this process recursively through the generation of a tree. In a decision tree, each leaf node receives a class label, non-terminal nodes, which include the root node and other internal nodes, contain attribute testing conditions to separate records that have different characteristics [16].

Figure 1 presents a typical decision tree schema. The X variables are the decision nodes. Each node is associated with an attribute, variables a and b represent the boundaries of the attributes that divide the decision into three tree paths, these can be nominal or numerical. The Class variables represent the leaves of the tree, which allow to classify the object under analysis.

In the late 1970s and early 1980s, J. Ross Quinlan developed ID3 (Iterative Dichotomiser), an algorithm to generate a decision tree. Some years after, he would propose the C4.5 algorithm, being an optimized version of the ID3. According to [3], it serves as a basis for new supervised methods.

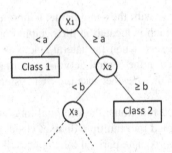

**Fig. 1.** Decision tree example.

The algorithm J48 is an extension of the C4.5 classification algorithm, arising from the need to recode into Java language, since C4.5 is originally written in the C language [12]. This algorithm uses the division and conquest method to increase the prediction capacity of decision trees. With this, it always uses the best locally evaluated step, without worrying if this step will produce the best solution, takes a problem and divides into several subproblems by creating sub-trees between the root and the leaves [2].

In general, the classification algorithms aim to search for models that reach the highest precision or the lowest error rate when applied to the test set.

## 3  Proposed Methodology

This paper aims to present a methodology for the use of decision trees as a means to understand brain activity from a case study with the recognition of spatial objects.

For the development of the case study the steps of the methodology proposed in Fig. 2 were applied. Hereafter, the definition of each item followed by the applications made in this work.

**Fig. 2.** Proposed methodology

1. **Hypothesis/Problem definition:** In this first step, we must formulate a problem or hypothesis to investigate its veracity, can be tested and confirmed or even refuted.

   Hypothesis: Sighted and visually impaired people use different areas of the brain to "visualize" space objects: sighted active people the occipital lobe and visually impaired people active the parietal lobe.

2. **Data collection:** The appropriate EEG equipment should be chosen for the collection of brain signals, as well as the necessary documentation required by the Research Ethics Committees.

The signals were recorded through Openvibe software, without the use of filters, and recorded directly in the GDF (Graphic Data Format) format. Steps to collect the brain signals:

- Tool for collection of brain signals: Actichamp with 32 channels.
- Software used to record brain signals: Openvibe.
- Protocol for the collections was as follows:
  - a protocol with general guidelines such as washing hair with neutral shampoo, not ingesting caffeine until 4 h prior to collection;
  - the collection was performed in a room, which was present the researcher, assistants and the participant of the experiment;
  - care was taken to ensure that there were no interruptions, noise pollution and unpleasant temperature;
  - with the equipment properly organized, the cap was inserted on the head of the individual;
  - the electrodes were stimulated until they had impedance low enough to perform the collection of brain signals;
  - a camera was positioned in front of the subject, in order to allow the collections to be recorded and later analyzed;
  - the individual was instructed to identify 3 geometric objects: cube, pyramid and parallelogram;

3. **Pre-processing:** The EEG recorded data must be processed by series of processes that allow data mining. The first step to be verified is the file format. After, we must extract the noises, adjusting the data according to the need of the application. Therefore, the filters were applied later, when the data was converted to CSV (Comma Separated Fields). For the data conversion, still in Openvibe, an algorithm was used to transform the GDF data into CSV and the application of filters. The data were treated to improve the reading of the signals.

- Cleaning of data, transformation of data into ARFF format (WEKA format).

4. **Applying decision tree:** After performing the previous steps, it is possible to execute the data mining algorithm. The algorithm will look for patterns using its strategies, through the data reported. To execute the chosen technique, we must choose the software for data mining, configure the parameters according we desire.

- Algorithm used: Decision algorithm J48
- Weka tool for data mining
- The algorithm was run from a set of data containing the highest values of each of the 32 channels of Actichamp. The data set contains the highest signal value every 100 ms. Tests with intervals of 25 and 50 instances were also performed.

5. **Analysis:** Finally, we analyze and validate the results obtained through the generated models. For this, it is important the participation of specialists in the area of study, in order to validate and guarantee the consistency of the results.

- Hypothesis confirmed for this study.

# 4  Case Study Using Spatial Activities

People without visual impairment, spatial analysis such as identification of objects or people, location and movement in space, is naturally dominated by the sense of sight. This sense is sent and processed in the occipital lobe. Therefore, people who are born without the vision or lose with the time, have this sense compromised, and they need to find other senses a way to overcome the obstacle of the absence of the vision [18].

[9] considers touch the most appropriate sense to provide the displacement references in space. This sense is processed in the parietal lobe.

According to [6]: "There is a common perceptual system to both the tactile and the visual modality: sighted person used a combination of modalities, but in a blind person they are accessible just in tactile sphere". In this way, the question has arisen: do sighted people and visually impaired use different brain areas when spatial ability is stimulated?

In this way the following hypothesis is adopted: Sighted people and visually impaired use different areas of the brain to "visualize" space objects. In sighted people, primarily the occipital lobe is activated. In the visually impaired, the parietal lobe is activated primarily.

## 4.1  Generated Models

From the application of the methodology proposed in Sect. 3, decision tree models were generated for each one of the four subjects. In subjects 1 and 2 are sighted and subjects 3 and 4 are congenital visually impaired.

### 4.1.1  Person 1

The channels that presented significant activity in the model presented in Fig. 3 were: TP10, CP6, FP1, F7, FP2, O1 and F3. The channels FP1, F7, FP2 and F3 correspond to the frontal lobe, as presented in Table 1, and it is responsible for the executive functions and management of cognitive resources; The TP10 channel belongs to the temporal lobe, which deals with the perception of biological movements; The CP6 channel, belongs to the parietal lobe and it is responsible for the tactile sense. The O1 channel belongs to the occipital lobe where the image processing takes place. For example, the form 1 (signed in Fig. 3 as number 1 in red) shows that TP10 and CP6 showed greater activity for the parallelogram and for the cube (number 2 in red).

### 4.1.2  Person 2

Figure 4 shows the relevant activity channels were: F3, OZ, T7, FT9 and TP9. The frontal lobe F3 and FT9, responsible for cognitive functions and reasoning; T7 and TP9 the temporal lobe, that is responsible for the perception of biological movements and the occipital lobe, OZ channel that involves visual perception, as well as the recognition of objects.

### 4.1.3  Person 3

The channels with relevant activity channels were: TP9 and T7, temporal lobe; FZ and F7, frontal lobe; and CP6, parietal lobe. The analysis performed through Table 1 shows

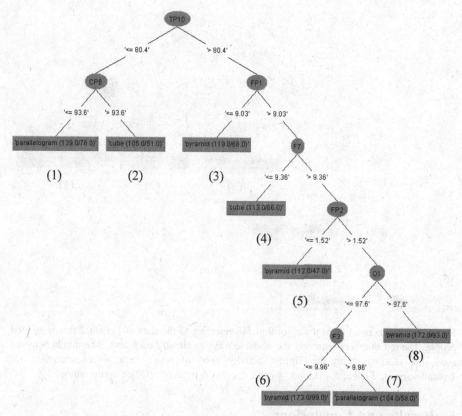

**Fig. 3.** Tree generated from the algorithm J48 referring to the data of Person 1 that is sighted people. The red numbers represent the possible ways to classify each task. The numbers in red show the channels with the highest activity used to reach each object according to: 1-parallelogram; 2-cube; 3-pyramid; 4-cube; 5-pyramid; 6-pyramid; 7-parallelogram; 8-pyramid. (Color figure online)

that the functions involved in object recognition were mainly: perception of biological movements (temporal lobe, TP9 and T7); Tactile sensation (parietal lobe, CP6); and the frontal area (FZ, F7), responsible for decision making and movement planning (Fig. 5).

### 4.1.4   Person 4
The generated tree of the subject 4 (Fig. 6), visually impaired, significant activity is observed in the channels O2 and P3. These channels involve two large areas, which is the occipital lobe (O2), responsible for the visual ability; and the parietal lobe (P3), which is used the tactile function.

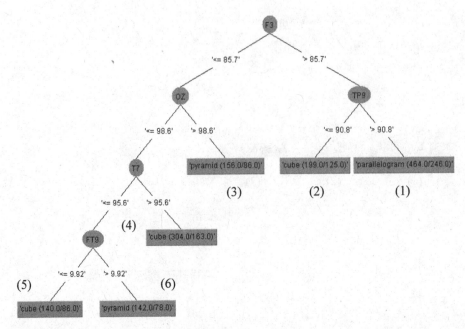

**Fig. 4.** Tree generated from the algorithm J48 referring to the data of Person 2 that is sighted people. The red numbers represent the possible ways to classify each task. The numbers in red show the channels with the highest activity used to reach each object according to: 1-parallelogram; 2-cube; 3-pyramid; 4-cube; 5-cube; 6-pyramid. (Color figure online)

## 5  Results and Conclusions

According to [6], there is a common perceptual system between the tactile and visual modality, in which individuals with normal vision can achieve insights from a combination of modalities that are accessible to the visually impaired only from the tactile sphere.

Therefore, the hypothesis of this work suggests that visually impaired and sighted people use different areas of the brain to "visualize" spatial objects. Sighted people use the occipital lobe, which is responsible for visualization, and visually impaired primarily use the parietal lobe, area responsible for tactile perception.

In order to verify the hypothesis in question, we performed tests using decision tree with the algorithm J48 in the data mining software Weka. The trees generated through the execution of the J48 algorithm did not present satisfactory accuracy rates for a classification task. However, the main idea of this work is not to classify the data, it is important to point out that for the analysis of the case study of this work the high rate was taken into account for the selection of the trees that were presented. However, the high rate was not determinant for the understanding of the electrode paths, since the intention was to analyze the generated trees visually to verify and understand the brain activities during the identification of the geometric objects.

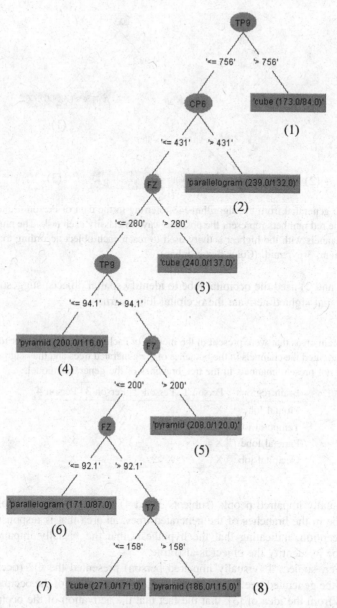

**Fig. 5.** Tree generated from the algorithm J48 referring to the data of Person 3 that is visually impaired. The red numbers represent the possible ways to classify each task. The numbers in red show the channels with the highest activity used to reach each object according to: 1-cube; 2-parallelogram; 3-cube, 4-pyramid; 5-pyramid; 6-parallelogram; 7-cube; 8-pyramid. (Color figure online)

Table 2 presents the summary of the areas that had some channel in the tree branches generated for each subject. Based on Table 2, we can see that sighted people

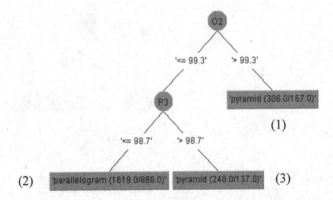

**Fig. 6.** Tree generated from the algorithm J48 referring to the data of Person 4 that is visually impaired. The red numbers represent the possible ways to classify each task. The numbers in red show the channels with the highest activity used to reach each object according to: 1-pyramid; 2-parallelogram; 3-pyramid. (Color figure online)

(subjects 1 and 2) used the occipital lobe to identify spatial objects, suggesting that the hypothesis that sighted activate the occipital lobe is true.

**Table 2.** Brain areas that were present in the models of each person. The X letter represents the areas that presented the channels in the branches of the generated trees and the - signal means that the area did not present channels in the tree branches of the generated models.

| Brain regions | Person 1 | Person 2 | Person 3 | Person 4 |
|---|---|---|---|---|
| Frontal lobe | X | X | X | – |
| Temporal lobe | X | X | X | – |
| Parietal lobe | X | – | X | X |
| Occipital lobe | X | X | – | X |

The visually impaired people (subjects 3 and 4) had electrodes that belong to the parietal lobe in the branches of the generated trees, an area that is responsible for the tactile perception, indicating that the hypothesis that the visually impaired use the parietal lobe to identify the object is also true.

However, subject 4 (visually impaired person) presented the O2 (occipital lobe) channel in the generated tree. The relevance in the brain signals of the occipital area can be derived from the idea of [3], that the fact that the activation of the occipital cortex has often indicated a reflection of the processes of mental visual images triggered by other modalities. In this way, the occipital lobe, during touch, may be the basis of cross-plasticity observed in congenital visually impaired.

In addition, we can see that visually impaired people use other abilities besides touch to do the recognition of objects, as affirms [18], that people who are born or lose the ability to see use of others senses to compensate for the absence of vision. Although the visually impaired and from sighted people presented positive results for the hypothesis, we believe that further works in this study, it is necessary more tests.

# References

1. Adamatti, D.F., Silveira, J., Carvalho, F.H.: Analyzing brain signals using decision trees: an approach based on neuroscience. Revista Eletrônica Argentina-Brasil de Tecnologias da Informação e da Comunicação **1**(5), 1–12 (2016)
2. Alvarenga, M.T.: Utilização da ferramenta j48 para descoberta do conhecimento em bases de dados fitossanitários, climáticos e espectrais. Master thesis, Universidade Federal de Lavras, Minas Gerais, Brazil (2014). (in Portugues)
3. Amedi, A., et al.: Cortical activity during tactile exploration of objects in blind and sighted humans. Restorative Neurol. Neurosci. **28**(2), 143–156 (2010)
4. Camilo, C.O., da Silva, J.C.: Mineração de dados: Conceitos, tarefas, métodos e ferramentas. Universidade Federal de Goiás (UFC), pp. 1–29 (2009)
5. Cosenza, R., Guerra, L.: Neurociência e educação. Artmed Editora (2011). (in Portugues)
6. Gardner, H.: Multiple Intelligences: New Horizons in Theory and Practice. Basic Books (2008)
7. Gazzaniga, M.S., Heatherton, T.F.: Psychological Science: Mind, Brain, and Behavior. Recording for the Blind & Dyslexic (2002)
8. Gil, M.: Deficiência Visual – Cadernos da TV Escola, n1/2000. MEC/Secretaria de Educação a distância. http://www.dominiopublico.gov.br/download/texto/me000344.pdf. Accessed 5 Oct 2015
9. Kastrup, V.: A invenção na ponta dos dedos: a reversão da atenção em pessoas com deficiência visual. Psicologia em Revista **13**(1), 69–90 (2007). (in Portuguese)
10. Lent, R.: Neurociência da mente e do comportamento. Grupo Gen-Guanabara Koogan (2000). (in Portuguese)
11. Lent, R.: Cem bilhões de neurônios: conceitos fundamentais de neurociência. Atheneu (2004)
12. Librelotto, S.R., Mozzaquatro, P.M.: Análise dos algoritmos de mineração J48 e Apriori aplicados na detecção de indicadores da qualidade de vida e saúde. Revista Interdisciplinar de Ensino, Pesquisa e Extensão **1**(1), 26–37 (2014). (in Portuguese)
13. Machado, S., et al.: Interface cérebro-computador: novas perspectivas para a reabilitação. Revista Neurociências **17**(4), 329–335 (2009). (in Portuguese)
14. da Silveira Nunes, S., Lomônaco, J.F.B.: Desenvolvimento de conceitos em cegos congênitos: caminhos de aquisição do conhecimento. Psicologia escolar e educacional **12** (1), 119–138 (2008). (in Portuguese)
15. Prada, B.M.L.: Interface cérebro-computador não invasiva baseada em OpenVibe. Mâster thesis, Universidade de Aveiro (2010)
16. Tan, P.-N., et al.: Introduction to Data Mining. Pearson Education India (2006)
17. Velloso, B.P., Pereira, A.T.C.: Sistema de monitoramento de atenção baseado em eletroencefalografia para avaliação de objetos de ensino e aprendizagem. Jornal Nuevas Ideas em Informática Educativa TISE **10**, 572–577 (2014)
18. de Viveiros, E.R.: Mindware semiótico-comunicativo: campos conceituais no ensino de física para deficientes visuais utilizando uma interface cérebro-computador. Ph.D. thesis, Universidade Estadual Paulista, 487-f (2013). (in Portuguese)

# Nonlinear Equations

# Multi-Network-Feedback-Error-Learning with Automatic Insertion: Validation to a Nonlinear System

Alex N.V. Santos, Paulo R.A. Ribeiro[✉], Areolino de Almeida Neto,
and Alexandre C.M. Oliveira

Universidade Federal do Maranhão, São Luis, MA, Brazil
alex.newman@lacmor.ufma.br, paulo.ribeiro@ecp.ufma.br,
{areolino,alexandre.cesar}@ufma.br

**Abstract.** Proportional-Integral-Derivative (PID) controller is broadly used to control industrial plants. However, intelligent control strategies are able to improve the performance of a PID controller. Some of them can do so without removing the current controller. One of these techniques is called Feedback-Error-Learning (FEL), which adds an artificial neural network (ANN) to the closed-loop system, alongside a PID controller, to improve the control. FEL strategy is inspired on a neurocomputacional model for control and learning of voluntary movement. Nevertheless, the introduced ANN may reach a local minimum and the performance of the control system may not improve anymore. Therefore, another ANN can be inserted to improve the control even further. Such a strategy is called Multi-Network-Feedback-Error-Learning (MNFEL) and is a FEL improvement that uses multiple ANNs instead of a single one. An automatic approach to insert new ANNs based on the standard deviation of the plant's error was proposed to substitute the manual insertion driven by a specialist. Even though interesting results with the automatic approach have been found, this method was applied to only one system that was the burner group of a Brazilian mining ore company. To validate the method on a different system, this work applies the MNFEL with automatic insertion to a cooling coil nonlinear system simulated by a Hammerstein-Wiener model. Results with this approach indicate improvement in both systems. The automatic approach reached smaller overshoot when compared with manual approach. Graphical analysis points out better performance achieved by FEL over PID-only and MNFEL over FEL.

**Keywords:** Neural networks · Feedback-Error-Learning · Multi-Network-Feedback-Error-Learning

## 1 Introduction

Proportional-Integral-Derivative (PID) controllers are commonly used in industry due to its simplicity and low effort design. The PID is a linear controller and requires the linearization of real control systems near a stable point to operate with some degree of

D.A.C. Barone et al. (Eds.): LAWCN 2017, CCIS 720, pp. 173–186, 2017.
https://doi.org/10.1007/978-3-319-71011-2_14

accuracy. Even after the linearization, the control can be very unpredictable. Artificial Intelligence techniques, like Fuzzy [1], are often employed to compensate for the nonlinearity.

Artificial Neural Networks (ANNs) are well known for its universal approximation capacity [2], fast adaptability and generic design with infinite possible connections between layers, being the perfect candidate for complicated nonlinear systems. Many researchers took advantage of those characteristics to successfully apply ANNs to adaptive nonlinear control systems [3–5]. One of the most common techniques is the Feedback-Error-Learning (FEL) [6–8]. FEL is a strategy that integrates an ANN alongside a working PID. The ANN will adjust the PID output compensating for the nonlinearity of the plant. The strong point of this technique is that no drastic change on the working control system has to be made. This characteristic makes the strategy more economically attractive than others and justifies the popularity amongst industries. Although being a very powerful method, defining the correct parameters in such a way that the most adequate approximation is found can be a demanding task. In most cases, the training phase has to be restarted multiple times, assigning the parameters values by trial and error.

One common technique to overcome this problem is to use multiple neural networks. The literature has been well supplied with different multi-network approaches. As example, researchers have developed multiple neural networks trained simultaneously with a switching mechanism that selects the model with the smallest error in the moment [9]. Another work uses multiple networks integrated to an AdaBoost algorithm, which will boost their training to get a good prediction control for a deep seabed-mining robot [10]. In both of these works, the presence of a coordinator is clear. In the first one, the switching mechanism has to make a decision on which model use based on a well-defined switching rule. In the latter, the AdaBoost algorithm is responsible for getting the last prediction function from the ensemble of networks.

The coordinator determines how the ANNs interact with each other, avoiding eventual conflicts. While it is easier to simply use multiple neural networks to find an approximation instead of using a single one, it can be very frustrating to make them work together.

Introduced by [11], the Multi-Network-Feedback-Error-Learning (MNFEL) is a strategy to integrate multiple ANNs based on FEL. MNFEL is considered simpler than others multiple neural networks approaches due to the absence of a coordinator, therefore using auto coordination. In the original work, a specialist is required to insert new ANNs to the process. The strategy has been successfully applied to both simulated and practical experiments achieving good results [12, 13]. An automatic insertion method of new networks was proposed in [14]. The method is based on the standard deviation of the error between the setpoint and the output of the plant.

The automatic insertion method performs well and has better control metrics than the manual approach [14]. However, previous paper applied this method to only one plant, without validation in a second system. This work extends the current literature on the subject by employing MNFEL with automatic insertion to a simulated cooling coil plant. Additionally, previously described experiments [14] on a neural model of a pelletizing plant burner group are re-conducted.

This paper is organized as follows: a brief FEL strategy explanation is found in Sect. 2. In Sect. 3 the MNFEL strategy is presented, followed by the technique of automatic insertion. Details about the simulation environments used for validating the automatic insertion are present in Sect. 4. In Sect. 5 the results are shown along with a brief discussion. Conclusions are drawn in Sect. 6.

## 2   FEL Strategy

The Feedback-Error-Learning (FEL) control strategy was introduced in [6–8] and uses an ANN as a feedforward controller alongside a Conventional Feedback Controller (CFC). The strategy is inspired on control and learning of voluntary movement on neuroscience studies about the central nervous system. The original work [6] points that the cerebrocerebellum acquires the inverse dynamics of the musculo-skeletal system while monitoring the trajectory and the motor command. The acquired inverse dynamics will then account for the motor command, substituting for other regions of the brain. In the FEL strategy, the ANN can acquire the inverse dynamics of the controlled system while the control is carried out by the CFC. As learning proceeded, the ANN will gradually substitute the CFC as the main controller [6].

One of the reasons that made this method so popular is that no huge change in operation has to be done in order to use it. Consider a control system in industry whose CFC is the three term Proportional-Integral-Derivative (PID) controller as an example. In order to apply the FEL strategy to deal with nonlinearities, the operational PID controller is kept and the ANN is placed in parallel, thus reducing upgrade costs and unavailability.

The basic scheme can be seen in Fig. 1. The error function is given by

$$E(z) = z^{-1}R(z) - G(U_{cfc}(z) + U_{ANN}(z))$$

where $E(z)$ is the error, $R(z)$ is the setpoint, $G(z)$ is the transfer function of the plant, $U_{cfc}$ is the CFC control action and $U_{ANN}$ is the ANN control action.

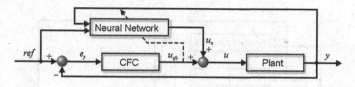

**Fig. 1.**  Basic FEL structure (from [11]).

The ANN is trained online via backpropagation using the CFC output as the error signal or teaching signal as it previously denoted. The CFC maintains the system stable until the ANN acquires the inverse dynamic model of the plant [15]. In [8], Gomi and Kawato proposed two different control approaches. The first one, Inverse Dynamic Model Learning (IDML) acquires the inverse dynamics of the system under consideration while the second, Nonlinear Regulator Learning (NRL), acts as a regulator for the PID, compensating for the nonlinearity.

Researchers have been developing the original literature ever since, presenting different approaches. As example, in [15] the high-order differentiators were replaced by a tapped delay line and the setpoint is delayed by m sampling periods.

## 3  MNFEL Strategy

Complex nonlinear systems can be approximated more efficiently using multiple neural networks rather than using a single ANN. The Multi-Network-Feedback-Error-Learning, MNFEL, is a FEL-based strategy that uses many ANNs cooperating with each other to acquire the inverse dynamic model of the plant or act as a nonlinearity regulator [11]. The cooperation is obtained by maintaining the ANN acquiring knowledge available to the others. Therefore, the MNFEL promotes the accumulative learning. This procedure, allied with the inclusion of new ANNs, can provide an escape from local minima [11, 12].

The ANN's training session is supervised by the specialist who will decide to insert a new network or maintain the current one training. The specialist's decision is based on graphical analysis and knowledge of the plant. The main idea is to train the ANN until the best performance is achieved. The synaptic weights will then be frozen and another ANN is inserted and trained. This process is repeated until the maximum number of networks is reached. Note that although some ANNs have finished their training phase, the MNFEL training will only be over when the last ANN finishes its training. The procedure is depicted in Fig. 2.

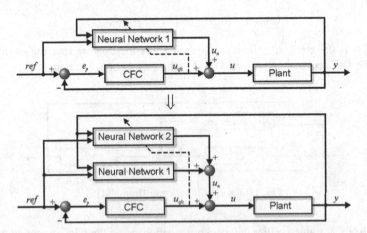

**Fig. 2.** MNFEL initial structure with a single ANN and after the insertion, with two ANNs (from [12]).

The MNFEL error function is

$$E(z) = z^{-1}R(z) - G(U_{cfc}(z) + \sum_{i=1}^{n} U_i(z))$$

where $n$ is the maximum number of networks and the ANNs outputs $U_i$ are accumulated and summed with $U_{cfc}$.

It is important to note (see Fig. 2) that the CFC's output used as error signal to train the first ANN is the same used to train the others. This does not mean that $u_{cfc}$ has the same value in different training sessions, even for the same setpoint (*ref*) [12].

Although in the original MNFEL a specialist is required to insert new ANNs, the strategy itself is structured in an auto-coordinated manner. In other words, it does not require a coordinator to combine and weight the outputs.

Previous works [11, 12] shown that a proper initialization of each ANN's weights in a way that its output is zero at the beginning of its training is necessary. That is because an ANN with random weights may produce bad responses, driving the system unstable. This initialization uses zero matrixes as initial weights between hidden and output layers, therefore does have a very low computational cost involved.

Actually, the strategy of training a single ANN at each time not only allows the accumulative learning but also takes less computational effort when compared with the simultaneous training of multiple networks.

## 3.1    Automatic Insertion of New ANNs

In the original MNFEL strategy, the specialist is responsible for introducing new ANNs into the system. That poses a huge drawback for practical applications, requiring a lot of effort and time. The automatic method of new ANNs insertion proposed in [14] is based on the standard deviation of the error of the plant. The process analyzes the error of the plant and identify whenever the dispersion has increased or is kept the same after a few training sessions. If the dispersion is increasing, that means the current configuration is struggling too much to approximate the inverse dynamic model. The standard deviation for the training session $p$ is given by

$$D_p = \sqrt{\frac{1}{N-1} \sum_{t=1}^{N} (e(t) - e_m)^2}$$

where $N$ is the number of epochs of each training session, $e(t)$ is the error of the plant at epoch $t$ and $e_m$ is the average of $e$.

# 4    Simulation Environment

## 4.1    Cooling Coil

A cooling coil is an important part of air conditioners and other systems where cooling is required. The cooling process happens when the contents, commonly air or water, get in contact with the surface of the coil as they move through. The cooling efficiency depends on the length of the coil.

How the contents are streamed through the coil depends on the content and the application. In this work a control valve $u$ is responsible for the air mass flow inside the coil, related to the coil outlet air temperature Y(s).

The cooling coil is simulated by a simplified Hammerstein-Wiener model [16]. Models as such represent the dynamic system as a series composition of static non-linearities and linear dynamics. The model equation is given in Eq. (1). The static nonlinearity is present in Eq. (2).

$$Y(s) = \mathcal{L}\{f(u(t))\}e^{-sT_d}\frac{1}{\tau s + 1} \tag{1}$$

$$f(u(t)) = \frac{1}{3.433}\ln(30u(t)+1) \tag{2}$$

The delay time of the plant is represented by $T_d$ and $\tau$ is a time constant associated with the cooling coil. Values used for $T_d$ and $\tau$ are 10 s and 200 s, respectively. The system dynamics can be approximated by the difference equation [17]:

$$y[n] = \left(1 - e^{\frac{-T_d}{\tau}}\right)f(u[n-3]) + e^{\frac{-T_d}{\tau}}y[n-1]$$

This nonlinear plant was chosen based on the simplicity and previous works involving FEL and its variations [18, 19].

## 4.2   Burner Group

This simulation environment is a neural model of a pelletizing plant's burner group. The model was developed by [20] to test the feasibility of both FEL and MNFEL strategies in controlling one burner group of a large Brazilian mining ore company's pelletizing plant.

The neural model is a MLP with a single hidden layer of 150 neurons, 10 input and 1 output neurons. The input is a tapped delay line with a history of the last 10 controller outputs, which are percentages of flow of oil to be burned. The output is a normalized value of temperature in °C respecting [−0.5, 0.5]. The actual interval of temperatures is [0, 1380].

In this simulation, the control system should be able to adjust the temperature with a low rise time but with a minimum overshoot, to prevent waste of oil. It is important to avoid reaching the maximum temperature of 1380 °C. On a real environment, that would destroy the burner's equipment.

# 5   Results

Different Multilayer Perceptrons (MLPs) networks and MNFEL configurations were tested in each one of the plants. Each experiment is replicated 10 times. The MLPs are single hidden layer without biases. The FEL approach NRL is used in the experiments described here. NRL was selected due to its better control behavior when compared with IDML in previous trials.

This work conducted extensive experiments to generate and test multiple parameters values used in each one of the systems. The learning rates are quite small, usually $10^{-5}$ or less. Small values for learning rates provide a fine adjustment for the synaptic weights and, therefore, have greater chance to achieve a good control effect without overshooting it.

Graphical analysis of each plant's control effect and root mean square error (RMSE) were used to estimate the number of both input and hidden neurons. The number of neurons in input layer and hidden layer are not the same for the two plants. In NRL approach, the input neurons refer to a tapped delay line of plant's error. If a small number of neurons are used in the input layer, a short history of errors is fed to the network. Naturally, different problems may require different ANN configurations to acquire an effective inverse dynamics model.

The analysis pointed that a MNFEL configuration with numerous input neurons is not suitable for the cooling coil plant. The history of errors in this case was too long and probably delayed the controller response. On the other hand, for the pelletizing plant, a greater number of input neurons showed a better control performance. The number of hidden neurons selected for each configuration and plant is based on RMSE and control effect observations.

The maximum number of neural networks is fixed for each plant based on the observed previous trials. Even though the method identifies a good time to insert new ANNs, it does not account for sufficiency. This means that the supplied number of ANNs will always be used regardless the reached control quality.

In order for the manual MNFEL experiments to be more precise, the specialist trained in each plant prior to the experiment. At the end of each training session, when the specialist had to choose between inserting a new network or keep on the current, the control was graphically analyzed alongside a few metrics like root mean square error (RMSE), rise time and percent overshoot.

Both systems under consideration behave as critically damped systems. Therefore the rise time metric is calculated between 5% and 95% of the setpoint. Since the range of values for the burner group plant is long and the static nonlinearity from this cooling coil Hammerstein-Wiener model requires values between 0 and 1, normalization is applied. All metrics were calculated from the normalized system outputs. The normalization method is described by

$$z_i = \frac{x_i - \min(x)}{\max(x) - \min(x)}$$

where $x$ is the set of numbers, being $x_i$ the $i$-th number while $z_i$ is the $i$-th normalized number.

## 5.1  Cooling Coil Experiments

Cooling coil is the device responsible for temperature control, thus oscillation is not desired. The PID gains are carefully selected in order to produce an initial stable system. The gains $K_p$, $K_i$, $K_d$ are set 2, 0.004 and 1, respectively. The ANNs with better performances for this experiment are small sized. Experiments on this plant indicate that a maximum of two ANNs is enough to produce an efficient control action.

Out of all experiments, the 4 configurations selected to be portrayed are present in Table 1. The ANNs are trained in the presented sequence. The Gamma I-H and gamma H-O columns refers to the learning rate used for the synaptic weights between the input and hidden layers; and hidden and output layers, respectively.

The system is simulated with a step response where the setpoint is 0.85, normalized between 0 and 1. The simulated time is 10000 s. However, all the experiments converged to the setpoint before reaching 5000 s.

**Table 1.** MNFEL configurations with 2 neural networks for cooling coil plant.

| Configuration | # ANN | Input neurons | Hidden neurons | Gamma I-H | Gamma H-O |
|---|---|---|---|---|---|
| CONFIG 1 | 1 | 10 | 12 | 0.00001 | 0.00001 |
|  | 2 | 3 | 2 | 0.00001 | 0.00003 |
| CONFIG 2 | 1 | 3 | 2 | 0.00001 | 0.00003 |
|  | 2 | 10 | 12 | 0.00001 | 0.00001 |
| CONFIG 3 | 1 | 10 | 12 | 0.00001 | 0.00001 |
|  | 2 | 4 | 3 | 0.000008 | 0.00001 |
| CONFIG 4 | 1 | 4 | 3 | 0.000008 | 0.00001 |
|  | 2 | 10 | 12 | 0.00001 | 0.00001 |

In Fig. 3 the systems output using auto MNFEL e manual MNFEL are compared. Response from both approaches is very similar but some differences are noticeable. The

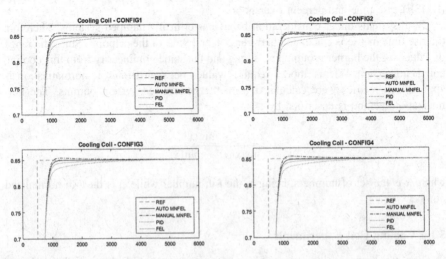

**Fig. 3.** Comparison between auto MNFEL, manual MNFEL, PID and FEL for cooling coil plant.

overshoot, for instance, is higher on the manual MNFEL. In this case, the rise time for the manual MNFEL is expected to be lower. Visually one can point that the auto MNFEL has a better control performance than manual MNFEL.

The performance of PID-only and FEL control is also depicted in Fig. 3. The conducted PID-only experiments used the conventional controller with previously described gains. The ANN used in each FEL experiment is the first network from its configuration (see Table 1). It is very noticeable that MNFEL strategy is able to improve the control achieved by FEL, while FEL performance is far better than the PID-only. This observation for all the configurations supports the arguments that FEL is able to provide a better control performance than a conventional method and that MNFEL is a FEL improvement by allowing a continuous learning when the previous ANNs reach the local minimum. For these experiments, the FEL training phases were carried out by the specialist.

The Table 2 translates in numbers the observed higher overshoot (see Fig. 3) for manual MNFEL. Observing the data is possible to realize that auto MNFEL had smaller overshoot and is more closely distributed over the experiments. For this kind of controlled system, both auto MNFEL and manual MNFEL have acceptable percentage overshoot.

**Table 2.** Percentage overshoot: average and standard deviation for both manual and auto MNFEL.

| Configuration | AVG % overshoot | | STDEV % overshoot | |
|---|---|---|---|---|
| | Auto | Manual | Auto | Manual |
| CONFIG 1 | 0.20244% | 1.12669% | 0.00261124 | 0.00361219 |
| CONFIG 2 | 0.36378% | 1.00205% | 0.00310599 | 0.00314560 |
| CONFIG 3 | 0.00000% | 0.53013% | 0.00000187 | 0.00302731 |
| CONFIG 4 | 0.08481% | 0.90744% | 0.00111014 | 0.00350598 |

On the other hand, the rise time from manual MNFEL is lower and do not deviate much from the average for multiple runs. The control metrics rise time and percentage overshoot are usually related, thus the expectation of a lower rise time when a higher percentage overshoot is present. The computed rise time is presented in Table 3.

**Table 3.** Rise time: average and standard deviation for both manual and auto MNFEL.

| Configuration | AVG rise time | | STDEV rise time | |
|---|---|---|---|---|
| | Auto | Manual | Auto | Manual |
| CONFIG 1 | 352.50 | 339.40 | 6.4226163 | 2.1071308 |
| CONFIG 2 | 348.30 | 338.70 | 7.6687678 | 2.8301943 |
| CONFIG 3 | 359.30 | 343.60 | 4.4056782 | 3.2310989 |
| CONFIG 4 | 351.80 | 340.10 | 4.7074409 | 3.4190642 |

It is important to mention that as percentage overshoot do not exceed 2% of the setpoint in any of the experiments, the settling time is assumed to be the same as the rise time.

In Table 4 the average RMSE and RMSE standard deviation for the experiments in each configuration are shown.

**Table 4.** Average RMSE and RMSE standard deviation for auto MNFEL and manual MNFEL.

| Configuration | AVG RMSE | | STDEV RMSE | |
|---|---|---|---|---|
| | Auto | Manual | Auto | Manual |
| CONFIG 1 | 0.0058031 | 0.0058000 | 0.00000517 | 0.00000444 |
| CONFIG 2 | 0.0058004 | 0.0057977 | 0.00000667 | 0.00000267 |
| CONFIG 3 | 0.0058094 | 0.0057968 | 0.00000514 | 0.00000089 |
| CONFIG 4 | 0.0058017 | 0.0057979 | 0.00000458 | 0.00000284 |

Although the smaller percent overshoot for the auto MNFEL, the RMSE metrics points in the opposite direction. The auto MNFEL configurations showed higher error and deviated more from the average. The difference is minimal and is almost unaccounted for when analyzed graphically (see Fig. 3). The RMSE, even though higher, indicates a successful automatic insertion. For the experimented configurations, the automatic method is able to introduce new ANNs with similar performance to supervised insertion.

## 5.2  Burner Group Experiments

The burner group neural model was developed to simulate control systems that aim to improve pellets quality and decrease oil consumption. Therefore, the minimum overshoot with maximum rise time possible is the desired scenario. The model is also constrained to the maximum temperature of 1380 °C which would cause the equipment to collapse and burn. For these experiments, the inputs and outputs are normalized with respect to the interval [−0.5 0.5].

The PID controller integrated to MNFEL is configured with gains 1.5, 0.01, 0.1 as $K_p$, $K_i$ and $K_d$, respectively. The setpoint is 1350 °C and simulation time is 1000 s. The system is stimulated with a step function. The neural networks with better performances for this plant usually have more input neurons than hidden neurons. The experiments presented in this work use MNFEL with 3 ANNs due to better results than 2-ANN MNFEL and 4-ANN MNFEL. The configurations are described in Table 5. The Gamma I-H and gamma H-O columns refers to the learning rate used for the synaptic weights between the input and hidden layers; and hidden and output layers, respectively.

A comparison between auto MNFEL and manual MNFEL in the burner group plant is depicted in Fig. 4. Like in the cooling coil experiments, the manual MNFEL visually overshoots more than the auto MNFEL. This metric is extremely important to evaluate the burner group control. Both approaches successfully have less than 1% overshoot.

**Table 5.** MNFEL configurations with 3 neural networks for burner group plant.

| Configuration | # ANN | Input neurons | Hidden neurons | Gamma I-H | Gamma H-O |
|---|---|---|---|---|---|
| CONFIG 1 | 1 | 20 | 15 | 0.00001 | 0.00005 |
|  | 2 | 40 | 15 | 0.000001 | 0.000005 |
|  | 3 | 55 | 15 | 0.00001 | 0.00005 |
| CONFIG 2 | 1 | 20 | 15 | 0.00001 | 0.00005 |
|  | 2 | 40 | 15 | 0.000001 | 0.000005 |
|  | 3 | 60 | 15 | 0.000008 | 0.000005 |
| CONFIG 3 | 1 | 60 | 15 | 0.000008 | 0.000005 |
|  | 2 | 20 | 15 | 0.00001 | 0.00005 |
|  | 3 | 40 | 15 | 0.000001 | 0.000005 |
| CONFIG 4 | 1 | 40 | 15 | 0.000001 | 0.000005 |
|  | 2 | 20 | 15 | 0.00001 | 0.00005 |
|  | 3 | 55 | 15 | 0.00001 | 0.00005 |

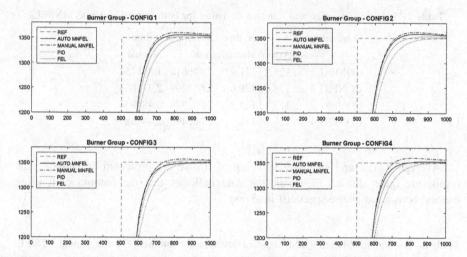

**Fig. 4.** Comparison between auto MNFEL, manual MNFEL, PID and FEL for burner group plant.

The controls produced by FEL and PID-only are also present in Fig. 4. The PID-only experiments used the same PID controller integrated to FEL and MNFEL experiments. The FEL experiments used the first neural network from its configuration, as described in Table 5. The graphical analysis leaves no doubt that FEL strategy performed better than PID-only. Meanwhile, the MNFEL control is observed to have a better rise time without large difference in overshoot when compared with FEL.

For these experiments, the FEL training phases were carried out by the specialist.

Table 6 shows the average and standard deviation for the percentage overshoot calculated for both experiments. As seen in Fig. 4, the manual MNFEL has superior percentage than auto MNFEL. The percentage overshoot value is also less sparse for auto MNFEL.

**Table 6.** Percentage overshoot: average and standard deviation for both manual and auto MNFEL.

| Configuration | AVG % overshoot | | STDEV % overshoot | |
|---|---|---|---|---|
| | Auto | Manual | Auto | Manual |
| CONFIG 1 | 0.00334% | 0.00521% | 0.00095220 | 0.00192470 |
| CONFIG 2 | 0.00354% | 0.00518% | 0.00139023 | 0.00226671 |
| CONFIG 3 | 0.00000% | 0.00475% | 0.00085044 | 0.00194732 |
| CONFIG 4 | 0.00112% | 0.00890% | 0.00059919 | 0.00311729 |

Table 7 depicts the rise time metric. Auto MNFEL and manual MNFEL are separated by a small difference but manual MNFEL, which has a higher overshoot, has lower rise time. The disparity between two approaches is minimal even when analyzing the standard deviation.

**Table 7.** Rise time: average and standard deviation for both manual and auto MNFEL.

| Configuration | AVG rise time | | STDEV rise time | |
|---|---|---|---|---|
| | Auto | Manual | Auto | Manual |
| CONFIG 1 | 125.3 | 121.9 | 1.846618 | 1.868154 |
| CONFIG 2 | 128.6 | 128.6 | 2.653299 | 2.374868 |
| CONFIG 3 | 125.9 | 125.2 | 2.022374 | 2.481934 |
| CONFIG 4 | 125.7 | 118 | 2.002498 | 2.449489 |

Average RMSE and standard deviation of RMSE are presented in Table 8. The results are quite similar. However, the data indicates that the automatic insertion method is as good as the specialist insertion.

**Table 8.** Average RMSE and RMSE standard deviation for auto MNFEL and manual MNFEL.

| Configuration | AVG RMSE | | STDEV RMSE | |
|---|---|---|---|---|
| | Auto | Manual | Auto | Manual |
| CONFIG 1 | 0.0062921 | 0.0062734 | 0.00001288 | 0.00001296 |
| CONFIG 2 | 0.0063187 | 0.0063203 | 0.00002087 | 0.00001806 |
| CONFIG 3 | 0.0063007 | 0.0062943 | 0.00001146 | 0.00001487 |
| CONFIG 4 | 0.0062959 | 0.0062600 | 0.00001395 | 0.00001070 |

# 6  Conclusion

This work applied the auto Multi-Network-Feedback-Error-Learning (MNFEL) to two different plants. A nonlinear Hammerstein-Wiener Model of a cooling coil and a neural model of a burner group from a pelletizing plant. The purpose was to validate the method of automatic insertion, checking whether or not it performs as good as a human insertion in more than one plant, extending the literature on the subject.

The method was not only able to efficiently control both tested plants but also capable of reducing the overshoot without negatively diminishing the rise time.

The results presented in this paper showed a better control performance when using Feedback-Error-Learning (FEL) strategy over the Proportional-Integral-Derivative (PID) controller. Even so, the performance achieved by MNFEL is superior when compared with FEL in both systems.

Multiple experiments with different configurations and its results serve as a strong base to support the argument that the method to automatically insert new neural networks to MNFEL is suitable to both plants and may as well be used in similar nonlinear plants.

The ongoing development on MNFEL involves the evaluation of this automatic insertion in heavily disturbed systems. Another future work is to increment the method to make it account for control sufficiency and reject new networks.

# References

1. Aafaque, M., Kadri, M.B.: Dynamic fuzzy modeling of cooling coil system. In: 17th International Multi-Topic Conference, INMIC, Pakistan, vol. 17, pp. 349–353. IEEE (2014)
2. Haykin, S.S.: Neural Networks: A Comprehensive Foundation, 2nd edn., USA (1998)
3. Singh, A.K., Gaur, P.: Adaptive control for non-linear systems using artificial neural network and its applications applied on inverted pendulum. In: India International Conference on Power Electronics 2010, IICPE 2010, India, vol. 1, pp. 1–8. IEEE (2010)
4. Qian, K., Chen, Z.: Dynamic inversion base on neural network applied to nonlinear flight control system. In: 2nd International Conference on Future Computer and Communication 2010, ICFCC, China, vol. 1, pp. 699–703. IEEE (2010)
5. Fu, M., Liu, T., Liu, J., Gao, S.: Neural network-based adaptive fast terminal sliding mode control for a class of SISO uncertain nonlinear systems. In: IEEE International Conference on Mechatronics and Automation, ICMA 2016, China, vol. 1, pp. 1456–1560. IEEE (2016)
6. Kawato, M., Furukawa, K., Suzuki, R.: A hierarquical neural-network model for control and learning of voluntary movement. Biol. Cybern. 57(3), 169–185 (1987)
7. Miyamoto, H., Kawato, M., Setoyama, T., Suzuki, R.: Feedback-error-learning neural networks for trajectory control of a robotic manipulator. Neural Netw. 1(3), 251–265 (1998)
8. Gomi, H., Kawato, M.: Learning control for a closed loop system using feedback-error-learning. In: 29th IEEE Conference on Decision and Control on Proceedings, United States, pp. 3289–2394. IEEE (1990)
9. Jia, C., Li, X., Liu, D., Ding, D.: Adaptive control using multiple parallel dynamic neural networks. In: 9th Asian Control Conference 2013, ASCC, Turkey, vol. 9, pp. 1–6. IEEE (2013)
10. Chen, F.: Predictive control based on multi-network for a deep seabed mining robot vehicle. In: 30th Chinese Control Conference 2011, CCC, China, vol. 30, pp. 2632–2635. IEEE (2011)
11. Almeida Neto, A.: Applications of multiple neural networks in mechatronics systems. Ph.D. thesis, Technological Institute of Aeronautics, Brazil (2003). (in Portuguese)
12. Almeida Neto, A., Goes, L.C.S., Nascimento, C.L.: Accumulative learning using multiple ANN for flexible link control. IEEE Trans. Aerosp. Eletronic Syst. 46(2), 508–524 (2010)

13. Ribeiro, P.R.A., Almeida Neto, A., Oliveira, A.C.M.: Multi-network-feedback-error-leaning in pelletizing plant control. In: 2nd International Conference on Advanced Computer Control, ICACC, China, vol. 2, pp. 340–344. IEEE (2010)

14. de Almeida Ribeiro, P.R., de Almeida Neto, A., de Oliveira, A.C.M.: Multi-network-feedback-error-learning with automatic insertion. In: Corchado, E., Novais, P., Analide, C., Sedano, J. (eds.) Soft Computing Models in Industrial and Environmental Applications, pp. 171–178. Springer, Heidelberg (2010). https://doi.org/10.1007/978-3-642-13161-5_22

15. Nascimento, C.L.: Artificial neural networks in control and optimization. Ph.D. thesis, University of Manchester, United Kingdom (1994)

16. Wang, S.: Dynamic simulation of building VAV air-conditioning system and evaluation of EMCS on-line control strategies. Build. Environ. **34**(6), 681–705 (1999)

17. Wu, Y., Dexter, A.L.: Modeling capabilities of fuzzy relational models. In: 12th IEEE International Conference on Fuzzy Systems, FUZZ 2003, USA, vol. 1, pp. 430–435. IEEE (2003)

18. Kadri, B.M.: Robust model free fuzzy adaptive controller with fuzzy and crisp feedback error learning schemes. In: 12th International Conference on Control, Automation and Systems, ICC, vol. 12, pp. 384–388. ICROS, Korea (2012)

19. Kadri, M.B., Dexter, A.L.: Disturbance rejection in information-poor systems using an adaptive model-free fuzzy controller. In: 28th North American Fuzzy Information Processing Society Annual Conference, NAFIPS 2009, USA, vol. 28, pp. 1–6. IEEE (2009)

20. Ribeiro, P.R.A., Costa, T.S., Barros, V.H., Almeida Neto, A., Oliveira, A.C.M.: Feedback-error-learning in pelletizing plant control. In: 7th Brazilian Meeting on Artificial Intelligence, ENIA, Brazil (2009)

# Nanoelectromechanical Systems

# Planned Obsolescence Using Nanotechnology for Protection Against Artificial Intelligence

Mirkos Ortiz Martins[1(✉)] and Dante Barone[2]

[1] Centro Universitário Franciscano, Santa Maria, RS, Brazil
mirkos@unifra.br
[2] Universidade Federal do Rio Grande do Sul, Porto Alegre, RS, Brazil
barone@inf.ufrgs.br

**Abstract.** One of the great ethical concerns of the study and development of artificial intelligence is the loss of control of machines, allowing them to overcome humanity in evolutionary terms and becoming hostile to the point of competing with and perhaps eliminating human beings. If an AI is truly conscious in its fullness, nothing prevents it from creating its own ethics and this characteristic is incompatible with human ethics. This work makes a theoretical questioning about the self-consciousness of an artificial intelligence, weighs its consequences and proposes a solution to an eventual conflict of species (machine versus human) using nanotechnology.

**Keywords:** Electronic design · Nanowire · Security development · Ethics

## 1  Introduction

Defining intelligence is not a simple task: it involves relating different areas of knowledge, creating hypotheses about the functioning of an organ still enveloped by a certain lack of knowledge - the brain. Could reproducing the brain function computationally be done without regard to consciousness?

In this line of the unknown, Roger Penrose in The Emperor's New Mind [1] describes human consciousness as non-algorithmic in its construction and not reproducible by computers obeying the architecture of a Turing machine. But, according to Marvin Minsky (1994): *"Will robots inherit the earth? Yes, but they will be our children"*.

But children with conscience or only actuators of reactions taught by algorithms that mimic a human brain? Minsky believed and argued that the reproduction of all the effectiveness of the human brain could be done through algorithms in machines, not differentiating in the end from each other (human and machine) [2]. This current of thought is known as Strong Artificial Intelligence (AI Strong), where a machine could reproduce feelings and have consciousness.

The independence between consciousness and intelligence is an important issue to be answered so that strong AI can actually be implemented. The technological domain for the construction of complex systems is already a fact, the

D.A.C. Barone et al. (Eds.): LAWCN 2017, CCIS 720, pp. 189–194, 2017.
https://doi.org/10.1007/978-3-319-71011-2_15

scale of material manipulation allows an integration in which thousands of circuits are added to a single chip so the development architecture imitating a brain is a matter of time, to be presented. The main question is whether this system would work according to the human brain.

Can AI systems without an associated consciousness be able to be classified as intelligent or can it only be described as an application reacting to external and environmental inputs? Developing an algorithm that emulates the consciousness is perhaps the holy grail of technological industry of the century. Then we will have a new format of intelligence living with humans and with an ethic of your own.

**Table 1.** Two possible scenarios for the coexistence in society of two intelligences: the human and machines with artificial intelligence. In (A) we see the two intelligences sharing the same set of ethical rules for peaceful coexistence, as in (B) each intelligence has its own ethics.

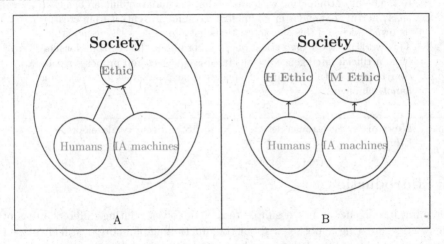

In a society with two intelligent species coexisting peacefully, the guarantee for the maintenance of peace between them is the use of a common ethic, with the convention of what is right, what is wrong and the active limits of action and thought. In the Table 1(A), a similar scenario is presented, with the human species and artificial intelligence sharing the same ethics.

In Table 1(B), the dangerous scenario represented by the fact that each species has its own set of ethical rules and these ethics are completely divergent makes humans and intelligent artificial systems competitive within a society. This competition can be by energy, civil and military dominance, by food, territory or dominance of knowledge. Any of these elements would build a brutal reality.

To make matters worse, humans with their mortality would be no match to artificial systems, which if they were really intelligent, would build their energy generators from inexhaustible sources such as solar, wind or nuclear fusion. These would be infinite systems.

One strategy for balancing forces in this unfavorable scenario for humans would be to limit the hardware life cycle where the system with artificial intelligence is running. One form of limitation is the use of planned obsolescence, suggested by this work.

## 2   Planned Obsolescence

The planned obsolescence is an industry practice for consumers to exchange their products in a given period of validity and thus continue consuming indefinitely. The practice of production with planned obsolescence was developed by North American industrial in the early 1900s and contribute to the life cycle of products reinforces the need for consumers continue to shop [3].

This is a pessimistic view of the use of programmed obsolescence. This paper proposes a discussion about the use of programmed obsolescence as a beneficial case for humanity as a tool for the defense of conscious intelligent systems that are executed in classical digital systems (based on silicon and carbon) with use of nanotechnology in design of hardware for artificial intelligence run. One of the ways to implement materials with programmed obsolescence is to manufacture them on a molecular scale, designing their functions with a limited number of repetitions in their use with the aid of nanotechnology.

### 2.1   Nanotechnology

Nanotechnology is the ability to manipulate materials on the nanometer scale, $10^{-9}$ m, where it is possible to work with molecules and atoms individually [4]. Currently, this technology has allowed the creation of increasingly efficient electronic devices with a smaller integration capacity of the order of 10 nm [5]. To get an idea, the size of a transistor in a next generation processor is about 1/10 the size of the bacterium *Mycoplasma pneumoniae* [6].

With the accuracy of developing hardware at the nanoscale, it is also possible to determine its behavior more precisely and it is in this sense that this work deals with the design of physical devices for the limitation, in the life cycle, for systems that perform their tasks with the use of artificial intelligence.

## 3   IA System with Physical Limits

There are studies on the behavior of atoms and molecules, when applied under voltage variations and evaluation of the time life and stability of certain configurations of nanometer scale material [7]. One of the devices used is called nanowire, a one-dimensional material with a thickness of a few atoms with lengths of tens to thousands of nanometers, exhibit aspect ratios (length-to-width ratio) of 100 or more [8].

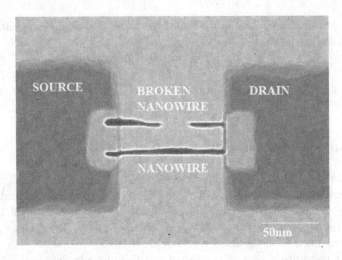

**Fig. 1.** Two nanowires of gold linking metal circuit. Above, a break nanowire after exceeding the maximum allowed voltage (Simulation of Electronic Microscopy, Author, 2016).

Nanowires have potential in the electronic field because there is almost electron-to-electron control [9] in the electronic transmission between their ends, which ensures a fine adjustment in electronic behavior in digital systems. An important item to observe on devices with nanowires is the limit of cycles that the component supports.

When a limit number of cycles of charge (or after exceeding the maximum allowed voltage) is reached by the nanowire, it breaks (Fig. 1), causing a closed-circuit effect or even conductance between the contacts previously connected by that nanowire [10].

The state of the art of electronic device construction makes massive use of redundancy both to provide fault tolerance and to prevent "death" of stretches of digital circuits in dedicated components such as microprocessors. Also regarding the construction of electronic devices, program logic can be physically inserted into a hardware through the use of FPGA (Field-Programmable Gate Array) [11], thereby the build of IA hardware is highly possible in highly specialized system [12].

The Table 2(A) shows a redundant circuit composed of Input, logic blocks containing hardware programming, three copies being in $Circuit_1$, $Circuit_2$ and $Circuit_3$. Finally, the last element of the schema is the Output. All elements are connected by lines, representing the nanowires. If the circuits contain logic to perform artificial intelligence, then the system has three hardware instances to avoid failures through redundancy, but if they are connected through nanowires, the limitation of the latter will determine the life cycle value of the digital circuit. This life cycle value is determined by nanoscale engineering, with variation of the nanowire material, thickness, number of elements making the connection between the Inputs, Circuits and Outputs.

**Table 2.** Circuit integrated with triple redundancy ($Circuit_1$, $Circuit_2$ and $Circuit_3$) connected from the input to the output through nanowires. In (A) the circuits are all functional while in (B) the nanowire left, from the input to the circuits, is broken, disabling the $Circuit_1$.

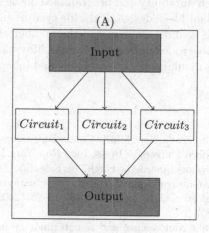

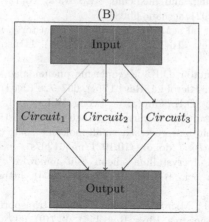

In the Table 2(B) it is possible to observe what would happen if the left nanowire, connecting Input and $Circuit_1$ breaks, making the circuit inoperative, but logic continues to work due to Redundancy. In order for the system to stop working completely, all three circuits must be stopped.

## 4    Conclusion

In systems running artificial intelligence, it is possible to model its execution in dedicated hardware, through programmed logic. To create a secure environment with intelligent systems, it would be important for the designer to worry about the independence of ethics that might eventually emerge from the machine. One way of protecting oneself is by limiting the life cycle of an intelligent system through planned obsolescence with the aid of nanotechnology engineering with the use of nanowires linking the logic components of integrated circuits.

The balance between durability and programmed obsolescence could be guaranteed through redundant block design. If the life cycle of an internal nanowire to the integrated circuit is well determined and the number of redundant blocks is implemented it is possible to formulate a maximum life estimate of the electronic component, protecting mankind about conscious and independent machines.

# References

1. Penrose, R.: The Emperor's New Mind: Concerning Computers, Minds, and the Laws of Physics. Oxford University Press, Inc., New York (1989)
2. Minsky, M.L.: Will robots inherit the earth? In: Scientific American, October 1994
3. Guiltnan, J.: Creative destruction and destructive creations: environmental ethics and planned obsolescence. J. Bus. Ethics **89**, 19–28 (2008). https://doi.org/10.1007/s10551-008-9907-9
4. Chan, W.W.C., et al.: Nanoscience and nanotechnology impacting diverse fields of science, engineering, and medicine. ACS Nano **10**(12), 10615–10617 (2016). https://doi.org/10.1021/acsnano.6b08335
5. Nourbakhsh, A., et al.: $MoS_2$ field-effect transistor with sub-10 nm channel length. Nano Lett. **16**(12), 7798–7806 (2016). https://doi.org/10.1021/acs.nanolett.6b03999
6. Waltes, K.B., Talkington, D.F.: Mycoplasma pneumoniae and its role as human pathogen. Rev. Clin. Microbiol. Rev. **17**(4), 697–728 (2004)
7. Kyaw, A.K.K., Wang, D.H., Gupta, V., Leong, W.L., Ke, L., Bazan, G.C., Heeger, A.J.: Intensity dependence of current-voltage characteristics and recombination in high-efficiency solution-processed small-molecule solar cells. ACS Nano **7**(5), 4569–4577 (2013). https://doi.org/10.1021/nn401267s
8. Wada, Y., et al.: Polycrystalline silicon "slit nanowire" for possible quantum devices. J. Vacuum Sci. Technol. B **12**, 48 (1994). https://doi.org/10.1116/1.587104
9. Tomczyk, M., et al.: Micrometer-scale ballistic transport of electron pairs in $LaAlO_3/SrTiO_3$ nanowires. Phys. Rev. Lett. **117**(9), 096801 (2016)
10. Kim, J.: Nickel silicide nanowire growth and applications. In: Lupu, N. (ed.) Nanowires Science and Technology. InTech (2010). https://doi.org/10.5772/39490
11. Anwer, J., Platzner, M.: Evaluating fault-tolerance of redundant FPGA structures using Boolean difference calculus. Microprocess. Microsyst. **52**, 160–172 (2017). https://doi.org/10.1016/j.micpro.2017.06.002
12. Tisan, A., Chin, J.: An end-user platform for FPGA-based design and rapid prototyping of feedforward artificial neural networks with on-chip backpropagation learning. IEEE Trans. Ind. Inform. **12**(3) (2016). https://doi.org/10.1109/TII.2016.2555936

# Author Index

Abdalla Júnior, Marcos Antônio   129
Adamatti, Diana Francisca   159
Aguilar, Wilbert G.   94
Alves, Lucas Ferreira   3
Andersson, Virginia O.   39, 81
Araujo Junior, Fernando Lopes   3
Araujo, Ricardo M.   39, 81
Araujo, Ricardo   31

Ballester, Pedro   31, 39
Bandeira, Vitor   17
Barone, Dante A.C.   63
Barone, Dante C.   143
Barone, Dante   189
Barroso, Márcio Falcão Santos   109, 129
Billa, Cleo Zanella   159
Birck, Marco A.F.   81
Birck, Marco F.   39
Birck, Marco   31
Bontorin, Guilherme   17

Cavalheiro, Esper Abrão   49
Chabert, Stéren   118
Cimini Júnior, Carlos Alberto   129
Cobeña, Bryan   94
Correa, Ulisses B.   31, 39

da Silveira Marciano, Jim Jones   109
de Almeida Neto, Areolino   173

de Almeida Ramos, Elias   17
dos Santos Bastos, Narúsci   159

Faber, Jean   49
Félix, Leonardo Bonato   129

Gomes, Rogerio Martins   3
Gonzalez, Rafael T.   63

Martins, Mirkos Ortiz   189
Mellado, Diego   118
Mendes, Eduardo Mazoni
      Andrade Marçal   109
Munsberg, Glauco R.   39

Nedel, Luciana P.   143

Oliveira, Alexandre C.M.   173

Reis, Ricardo   17
Riascos, Jaime A.   63, 143
Ribeiro, Paulo R.A.   173

Saavedra, Carolina   118
Salas, Rodrigo   118
Salcedo, Vinicio S.   94
Sandoval, David S.   94
Santos, Alex N.V.   173
Santos, Bruno Andre   3

Via, Guillem   49

Printed in the United States
By Bookmasters